Faouzi Derbel, Nabil Derbel and Olfa Kanoun (Eds.)
Power Systems & Smart Energies

Advances in Systems, Signals and Devices

—

Edited by
Olfa Kanoun, University of Chemnitz, Germany

Volume 7

Power Systems & Smart Energies

Edited by
Faouzi Derbel, Nabil Derbel and Olfa Kanoun

DE GRUYTER
OLDENBOURG

Editors of this Volume

Prof. Dr.-Ing. Faouzi Derbel
Leipzig University of Applied Sciences
Chair of Smart Diagnostic and Online Monitoring
Wächterstrasse 13
04107 Leipzig, Germany
faouzi.derbel@htwk-leipzig.de

Prof. Dr.-Ing. Olfa Kanoun
Technische Universität Chemnitz
Chair of Measurement and Sensor Technology
Reichenhainer Strasse 70
09126 Chemnitz, Germany
olfa.kanoun@etit.tu-chemnitz.de

Prof. Dr.-Eng. Nabil Derbel
University of Sfax
Sfax National Engineering School
Control & Energy Management Laboratory
1173 BP, 3038 SFAX, Tunisia
n.derbel@enis.rnu.tn

ISBN 978-3-11-046820-5
e-ISBN (PDF) 978-3-11-047052-9
e-ISBN (EPUB) 978-3-11-044628-9
ISSN 2365-7493

Library of Congress Control Number: 2018934305

Bibliographic information published by the Deutsche Nationalbibliothek
The Deutsche Nationalbibliothek lists this publication in the Deutsche Nationalbibliografie;
detailed bibliographic data are available on the Internet at http://dnb.dnb.de.

© 2018 Walter de Gruyter GmbH, Berlin/Boston
Typesetting: Konvertus, Haarlem
Printing and binding: CPI books GmbH, Leck

www.degruyter.com

Preface of the Editors

The seventh volume of the Serie "Advances in Systems, Signals and Devices" (**ASSD**), is a peer reviewed international scientific volume devoted to the field of power systems and smart energies. The series encompasses all aspects of research, development and applications of the science and technology in these fields. The topics concern Energy systems and energy transmission, renewable energy systems, hybrid renewable energy systems, photovoltaic systems, solar energy, wind energy, energy storage, batteries, thermal energy, combined heat and thermal power generation, electric machine design, electric machines modeling and control, electrical vehicles, technologies for electro mobility, special machines, variable speed drives, variable speed generating systems, automotive electrical systems, monitoring and diagnostics, electromagnetic compatibility, power systems, transformers, power electronics, topologies and control of power electronic converters.

Every issue is edited by a special editorial board including renowned scientist from all over the world.

Authors are encouraged to submit novel contributions which include results of research or experimental work discussing new developments in the field of power systems and smart energies. The series can be also addressed for editing special issues for novel developments in specific fields. Guest editors are encouraged to make proposals to the editor in chief of the corresponding main field.

The aim of this international series is to promote the international scientific progress in the fields of systems, signals and Devices.

It is a big pleasure of ours to work together with the international editorial board consisting of renowned scientists in the field of power systems and smart energies.

Editors-in-Chief
Faouzi Derben, Nabil Derbel and Olfa Kanoun

De Gruyter Oldenbourg, ASSD – Advances in Systems, Signals and Devices, Volume 7, 2018, p. V.
https://doi.org/10.1515/9783110470529-202

Advances in Systems, Signals and Devices

Series Editor:

Prof. Dr.-Ing. Olfa Kanoun
Technische Universität Chemnitz, Germany.
olfa.kanoun@etit.tu-chemnitz.de

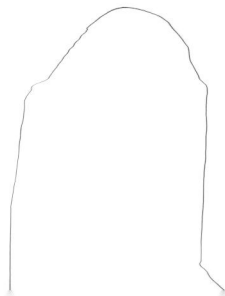

Editors in Chief:

Systems, Automation & Control

Prof. Dr.-Eng. Nabil Derbel
ENIS, University of Sfax, Tunisia
n.derbel@enis.rnu.tn

Power Systems & Smart Energies

Prof. Dr.-Ing. Faouzi Derbel
Leipzig Univ. of Applied Sciences, Germany
faouzi.derbel@htwk-leipzig.de

Communication, Signal Processing & Information Technology

Prof. Dr.-Ing. Faouzi Derbel
Leipzig Univ. of Applied Sciences, Germany
faouzi.derbel@htwk-leipzig.de

Sensors, Circuits & Instrumentation Systems

Prof. Dr.-Ing. Olfa Kanoun
Technische Universität Chemnitz, Germany
olfa.kanoun@etit.tu-chemnitz.de

Communication, Signal Processing & Information Technology

Sensors, Circuits & Instrumentation Systems

Advances in Systems, Signals and Devices

Volume 1
N. Derbel (Ed.)
Systems, Automation, and Control, 2016
ISBN 978-3-11-044376-9, e-ISBN 978-3-11-044843-6, e-ISBN (EPUB) 978-3-11-044627-2, Set-ISBN 978-3-11-044844-3

Volume 2
O. Kanoun, F. Derbel, N. Derbel (Eds.)
Sensors, Circuits and Instrumentation Systems, 2016
ISBN 978-3-11-046819-9, e-ISBN 978-3-11-047044-4, e-ISBN (EPUB) 978-3-11-046849-6, Set-ISBN 978-3-11-047045-1

Volume 3
F. Derbel, N. Derbel, O. Kanoun (Eds.)
Power Systems & Smart Energies, 2016
ISBN 978-3-11-044615-9, e-ISBN 978-3-11-044841-2, e-ISBN (EPUB) 978-3-11-044628-9, Set-ISBN 978-3-11-044842-9

Volume 4
F. Derbel, N. Derbel, O. Kanoun (Eds.)
Communication, Signal Processing & Information Technology, 2016
ISBN 978-3-11-044616-6, e-ISBN 978-3-11-044839-9, e-ISBN (EPUB) 978-3-11-043618-1, Set-ISBN 978-3-11-044840-5

Volume 5
N. Derbel, F. Derbel, O. Kanoun (Eds.)
Systems, Automation, and Control, 2017
ISBN 978-3-11-046821-2, e-ISBN 978-3-11-047046-8, e-ISBN (EPUB) 978-3-11-046850-2, Set-ISBN 978-3-11-047047-5

Volume 6
O. Kanoun, F. Derbel, N. Derbel (Eds.)
Sensors, Circuits and Instrumentation Systems, 2017
ISBN 978-3-11-044619-7, e-ISBN 978-3-11-044837-5, e-ISBN (EPUB) 978-3-11-044624-1, Set-ISBN 978-3-11-044838-2

Volume 7
F. Derbel, N. Derbel, O. Kanoun (Eds.)
Power Systems & Smart Energies, 2017
ISBN 978-3-11-046820-5, e-ISBN 978-3-11-047052-9, e-ISBN (EPUB) 978-3-11-044628-9, Set-ISBN 978-3-11-047053-6

Volume 8
F. Derbel, O. Kanoun, N. Derbel (Eds.)
Communication, Signal Processing & Information Technology, 2017
ISBN 978-3-11-046822-9, e-ISBN 978-3-11-047038-3, e-ISBN (EPUB) 978-3-11-046841-0, Set-ISBN 978-3-11-047039-0

Contents

K. Keitsch and T. Bruckner

Improving Short Term Load Forecasting with a Novel Hybrid Model Approach as a Precondition for Algorithmic Trading

Abstract: Considering the changing European power market, accurate electric load forecasts are of significant importance for power traders to reduce costs for ancillary services by leveling their position on continuous intraday power markets. The first part of the following case study, based on publicly available load data, focuses on a novel approach to combine different forecasting methodologies and techniques from the area of computational intelligence. The proposed hybrid model blends input forecasts from artificial neuronal networks, multi variable linear regression and support vector regression machine models with fuzzy sets to intraday and day-ahead forecasts. The forecasts are evaluated with commonly used metrics (mean average percentage error - MAPE & normalized rooted mean square error - NRMSE) to allow comparisons with other case studies.

Results from input forecasting models range from a yearly MAPE of 2.36 % for the linear regression model to 2.1 % for the support vector machine. Blended forecast from proposed hybrid models results in a MAPE of 1.46 % for one hour and a MAPE of 1.72 % for 24 hours ahead forecasts.

In the outlook section of the paper we show how to use blended forecasts as an input source for automatized intraday power trading with algorithms. Besides outlining use cases, model structure of trading algorithms and back testing approaches, the paper offers a state of the art insight on algorithmic trading in the power industry.

Keywords: Hybrid Forecast Model, Artificial Neuronal Network, Fuzzy Composition, Support Vector Machine for Regression, Algorithmic Trading

1 Introduction

The European power market is changing. While the integration of national markets into an European power market continues [1], the German Federal Ministry for Economic Affairs and Energy (BMWi) has published a White Book recently containing a new power market design including measures to support the German transition

K. Keitsch and T. Bruckner: K. Keitsch, Energy and Risk Management business unit at EXXETA AG in Karlsruhe, Germany, email: krischan.keitsch@exxeta.com, T. Bruckner, Institute for Infrastructure and Resources Management, Chair for Energy Management and Sustainability, University of Leipzig, Germany, email: bruckner@wifa.uni-leipzig.de

De Gruyter Oldenbourg, ASSD – Advances in Systems, Signals and Devices, Volume 7, 2018, pp. 1–18.
https://doi.org/10.1515/9783110470529-001

called "Energiewende" [2]. These measures correspond to the increasing shares of re-
newables, for instance 38 GW solar and 35 GW wind compared to 77.5 GW conventional
power plants (natural gas, lignite and hard coal) and remaining 12 GW nuclear power
plants [3], as well as changing demand patterns from residential households.

The White Book suggests measures in order to increase balance area loyalty
("Bilanzkreistreue"). The German power market is organized in several balancing
areas and every power utility was obliged to aim at a balanced area in order to
level demand and production. Grid operators handle discrepancies of production and
demand by utilizing reserve power. Resulting costs are billed to the power utilities.
A preferred approach to reduce balancing costs is to continuously trade on European
Intraday power markets such as EPEXSpot[1] and Nordpool Spot.[2]

The energy system transition challenges traditional business models of power
utilities [4] resulting in cost cutting programs which reduce, for instance, balance area
accounting costs. Thus, accurate short term load forecasting (STLF) for algorithmic
trading is vital for power traders, grid operators and power plant dispatch.

This paper aims at suggesting an STLF model utilizing and combining different
methodologies for generating precise electrical load prognoses. The hybrid model is
trained and tested on publicly available electric load data. Figure 1 represents German
load data for the year 2014. This paper analyses and verifies the following conjectures:

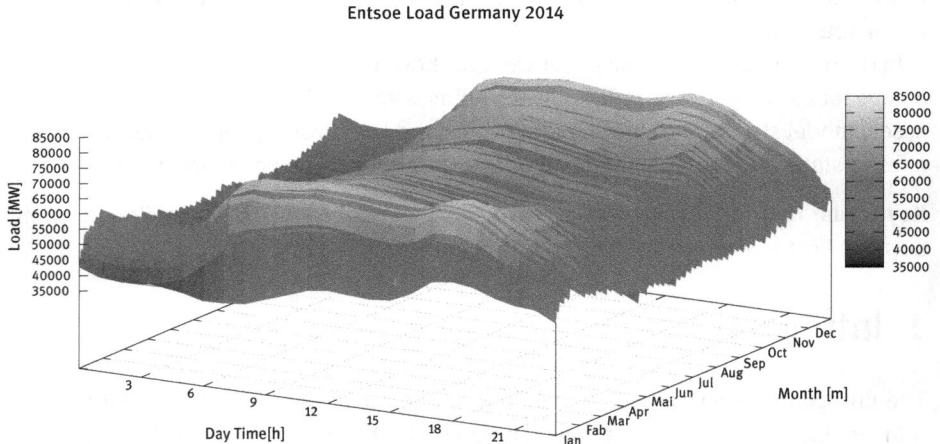

Fig. 1. Visualization of electrical load of Germany in 2014. Data source: ENTSO-E.

---

1 http://www.epexspot.com/
2 http://www.nordpoolspot.com/

Conjecture 1: Combining STLF model methodologies in a hybrid model framework increases forecasting accuracy of blends compared to the sole utilization of STLF input models.

The second hypothesis focuses on the contribution of input forecasts to blended prognoses:

Conjecture 2: Input forecasts with mediocre evaluation metrics results significantly contribute to the blended prognoses.

The author organized his paper as follows: In section 2 a brief literature review focuses on short term load forecasting with hybrid models and combining forecasts. The methodology of the novel short term load forecasting approach is represented in section 3, followed by Results, section 4, and Conclusion, section 5. The paper closes with an outlook on continuously refined forecasts being used by Algorithmic Trading Strategies in order to close resulting open intraday positions on intraday markets in section 6.

2 Literature review

Forecasting electrical demand is a domain of artificial neuronal networks (ANN), regression methodologies and support vector machines (SVM). A recent trend are so called hybrid models for STLF [5, 6]. Hybrid models combine the mentioned methodologies with techniques from fields such as computational intelligence. These techniques might include fuzzy logic, evolutionary algorithms, expert systems or wavelet transformation. The following section will discuss a selection of recently published papers focusing on hybrid models.

Hybrid Models: Combinations of fuzzy logic and ANNs are used in several publications. A fuzzy inference system (FIS) to classify input data [7] and a fuzzy expert system (FES) [8] to perform adaption to forecast time series are used to increase accuracy for instance. Adaptive Neuro-Fuzzy inference systems (ANFIS) are used in [9]. In [10] an advanced learning algorithm is applied to a neuro fuzzy network (NFN) and compared to an ANFIS model. An extended ANFIS model is presented in [11], while a combination of an ANFIS model and a genetic algorithm (GA) is proposed in [12] in order to select efficient training data sets. A GA is also used in [13, 14] to optimize the weights of an ANN. Results are compared to an ANN trained with the Levenberg-Marquardt (LM) algorithm.

An input data decomposition with a wavelet transformations (WT) approach is described in [15–17]. The WT technique splits load curve time series into subcomponents. Each subcomponent is then forecasted by a different ANN model.

The technique particle swarm optimization (PSO) is used in [18] to optimize weights of an ANN while in [19, 20] PSO is applied to SVR calibration to increase overall forecasting accuracy. In [21] a self organizing map (SOM) is applied to load data sets in order to perform a clustering. The identified input data clusters are then forwarded to individual SVR models. Data decomposition by clustering is also applied in [22].

The mentioned hybrid models for STLF apply, e. g., evolutionary algorithms and PSO to optimize forecasting methodologies such as ANNs and SVM for regression. Data decomposition is performed by clustering techniques such as SOMs or fuzzy logic sets and wavelet transformations. Adjustment to forecasts generated may be performed by FES.

Combining forecasts: The combination of forecasts may increase accuracy for intraday and day ahead forecasts. One example is represented in [23]. The authors suggest a case study based on results of their STLF model from 20 utilities in the United States. Their model consists of ANN based modules to forecast weakly (WM), daily (DM) and hourly (HM) load. The different load forecasts are combined according to the following approach: $L(k) = \alpha_{WM}(k)L_{WM}(k) + \alpha_{DM}(k)L_{DM}(k) + \alpha_{HM}(k)L_{HM}(k)$ for $k = 1..24$. $L(k)$ being the module load forecasts and α the weighting factor. The weighting factor α is determined by a recursive least square optimization analyzing the sum of square errors from actual load and forecasts. Results are represented for varying time ranges from 3 to 12 months for different utilities with an average MAPE of 2.34 %.

Load data from New England, United States, with a maximum power consumption of 27 GW is used in a case study [17]. The authors use a wavelet approach to decompose load data time series into sub data sets. Each sub set is used to train separate ANN. The ANN generates forecasts which are combined by a weighting factor to build the resulting overall forecast. The weighting factor is determined by a partial least square approach. Based on five years of training data results are depicted for a one year period ranging from 0.35 % (+1 h) to 1.84 % (+24 h) MAPE. The authors point out that their proposed hybrid model exceeds other models by more than 60 % (+1 h) to 23 % (+24 h).

A case study based on two data sets comparing complex weighting and aggregation rules is presented in [24]. The first data set is located in Slovakia. 35 models forecast the load of one hour per day. These models called "experts" are not described and are considered as black boxes. The use of non-normalized evaluation metrics interferes with any performance comparison. The second data set is based on load data from France. In this case 24 expert models (mainly regression models with different configurations) perform STLF for 320 days a year as holidays are excluded. Again complex strategies are used to determine weighting factors for expert forecasts provided. The authors report a 5 % and 15 % increase of forecasting accuracy for the Slovakian and the French data set.

3 Methodology

This section describes the methodology of the proposed hybrid forecasting model. This approach differs from hybrid models presented in the literature section. While forecasting models are combined with additional techniques, the hybrid model in this paper combines different forecasting models and utilizes additional techniques to improve forecasting accuracy. Another advantage of our proposed methodology is its modularity. Model cores may be easily replaced by more advanced models as development continues.

The proposed approach works as follows for model calibration Fig. (2): First, the available input data (electrical load, weather and economic data) is decomposed (section 3.1) based on calendar information (hierarchical clustering). Second, for each input data sub set the STLF models (section 3.2) are calibrated. All model cores for data sub sets build the model array for the forecasting process.

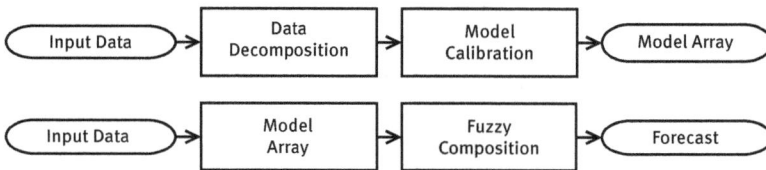

Fig. 2. Visualization of model calibration and forecasting process.

The forecasting process is illustrated on the bottom side of Fig. 2: First, input parameters based on calendar information (weather and economic data) are fed into the forecasting model array. As a second step, model array results are combined into forecast time series. Forecast composition works with fuzzy sets in order to smoothen forecast time series (section 3.3). As a third step, the prognoses from different model arrays are blended (section 3.4) based on an ex-post analysis.

3.1 Decomposition

This case study uses electrical load data provided by the European Network of Transmission System Operators (ENTSO-E)[3] for Germany. It is defined as follows:

> Actual [...] total load being defined as equal to the sum of power generated by plants on both TSO/DSO networks, from which is deduced: - the balance (export-import) of exchanges on interconnections between neighbouring bidding zones. - the power absorbed by energy storage resources. [25, p. 11]

---

3 https://www.entsoe.eu

The STLF model cores are calibrated with data ranging from 2006 until 2013. Data collected in 2014 is then used to validate the models. The load data used in this case study is processed according to [26].

Weather data from the German Meteorological Service (DWD)[4] for major German cities such as Berlin, Hamburg, Cologne, Munich and Stuttgart is applied as model input. Besides the air temperature and global radiation additional data such as precipitation, cloudiness, wind speed and humidity is available. However, sensitivity analysis revealed significant impact on forecasting quality of only a few parameters as shown in Tab. 1. In addition to the weather data the economic production index [27] takes into account the load's overall dependence on the economic growth rate and the rising production efficiency. The source of the production index is the Federal Statistical Office (Destatis).[5]

Tab. 1. Listing of relevant input data and training data set size for model cores.

Model	training Data Sets	Temp. Index	Cloud Index	Prod. Index
ANN	2010–2013	yes	yes	yes
TYPE	2010–2013	no	no	no
REG	2006–2013	yes	no	no
MVLR	2008–2013	yes	no	yes
SVR	2006–2013	yes	no	yes

Load and input data referred to is decomposed into separate sub data sets based on calendar information. The data sets represent four seasons, and each day of the week. National holidays are treated separately.

3.2 Model cores

Different methodologies are used as forecasting model cores.[6] The following section briefly describes our forecasting models:

The first model core is an ANN (*ANN-Model*). The ANN is trained with temperature, cloud and production index as input parameters. The setup of the ANN is determined by suggestions from [28] resulting in an MLP with three neurons in one hidden layer. Each ANN model core of the model array is trained with the LM algorithm. Due to the simple topology of the ANN models training requires less computational effort.

---

4 http://www.dwd.de

5 https://www.destatis.de/

6 The models are implemented in octave https://www.gnu.org/software/octave/.

The second model is based on type days from the training data. Each model represents the average load of a certain type day of one of the four seasons. The type day model uses no additional data as input and is called *TYPE-Model*.

The next two models perform a linear regression to learn the correlation between electrical load and input data. The first regression models depend on air temperature only (*REG-Model*) while the second regression model depends on air temperature and production index. This multi variable linear regression model is called *MVLR-Model*.

The last model core uses SVM for regression (SVR) and is based on [29]. The SVR models' parameters are determined by a grid search as suggested in [30]. The SVR model uses temperature and economic data as input and is called *SVR-Model*.

Forecasting model cores described above are used as input forecasts for the blending (section 3.4) process of this proposed hybrid model approach.

3.3 Fuzzy-composition

For higher forecasting accuracy individual forecasts from the model array are composed with fuzzy sets as shown in Fig. 3. Without fuzzy sets the model cores would generate a prognosis for their respective season only. The forecast for the next season would be made by a different model core. This leads to inaccuracies as seasons change. Hence, the fuzzy sets allow forecasts based on two model cores. A model core will generate forecasts for its season with fuzzy edges leading to prognoses with, for instance, a little bit of summer and a greater portion of spring. This technique helps to reduce the impact of a model change from one season to another. In this case study the linear fuzzy sets showed the best improvements. The fuzzy sets of the seasons overlap four weeks.

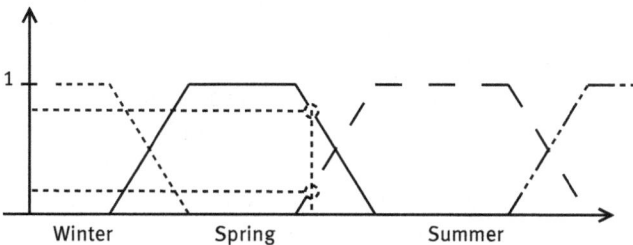

Fig. 3. Visualization of fuzzy sets to compose forecast time series from model arrays.

3.4 Blending STLF models

Another aspect of this proposed STLF hybrid model is blending available forecasts (input forecasts) from the model arrays. Based on an ex-post analysis at t_0 the best forecast or the best blend of two forecasts is determined. This ex-post information is then used to select or to blend an intraday or day ahead forecast based on the available prognoses of the different model arrays.

The ex-post analysis is of central importance in this process. Figure 4 visualizes three different cases:

Case 1: All input forecasts at t_{-1} are above the actual load. Input forecast closest to the actual load delivered the best forecasting results (prognosis p_2 as shown in the top Fig. 4).

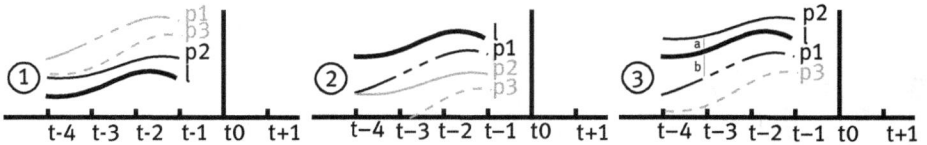

Fig. 4. Exemplary visualization of the three cases for the ex-post analysis.

Case 2: Analogue to case 1. All input forecasts underestimated the actual load and are positioned below the actual load (middle Fig. 4). The best forecast is p_1 in this example.

Case 3: The actual load is in between two input forecasts. In this example, prognoses p_2 and p_1 cover the actual load as shown in the bottom Fig. 4. Other available forecasts such as p_3 in this case may be ignored, as they have the lowest accuracy in this example.

As an example, at t_{-1} in case 3, the optimal weight of prognoses p_1 and p_2 is determined by deltas a and b of prognoses p_1 and p_2 to the actual load l. The optimal weighted prognosis p_w would have been $p_w = \frac{a}{a+b} \cdot p_1 + \frac{b}{a+b} \cdot p_2$.

In addition to selecting the best fitting input forecast in case 1 and 2 the total distance to the ex-post load the input forecast is determined and used as level adjustment information. The blended intraday forecast for one hour ahead will be called *Blend 1h*. A composed day ahead forecast will be called *Blend 24h*. While we assume that for blended intraday forecasts current load information is available, we make the following assumption for the day ahead forecast: Due to different market regimes, intraday markets allow a continuous trading with gates closed 30 min before delivery, the day head position has to be cleared with an auction. Gate closure is at 12 o'clock, delivery begins the next day at 0:00 o'clock. This leaves a gap of 12 hours. In our day ahead forecast we set a 14 h offset, including 12 hours gap till delivery and

2 hours for processing (analyzing ex-post performance of input forecasts, generating blended day ahead forecast and creating and uploading the spot market sheet). In conclusion our day ahead blended forecast is based on 14 hour old input forecasts.

The second strategy in blending algorithm is timely weighting of input forecasts. In Fig. 5, four different approaches of timely weighting the ex-post information are shown.

Linear: The top left plot symbolizes a linear weighting of the ex-post information of all three cases. The influence of the ex-post weighting information decreases by aging, as shown in (1).

Constant: The top right plot takes the ex-post information equally into account. Information from the near past and the remote past is treated the same, as shown in (2).

Progressive: In the bottom left plot ex-post information of the near past has a higher influence on weighting than information from the remote past, as shown in (3).

Regressive: Information of the past loses its influence on a regressive path as shown in bottom right plot in Fig. 5. This means that near and middle past information is taken into account for the weighted forecasts while the remote past has less influence, as shown in (4).

The time range of the past to learn from (Analysis Time Range - ATR) is determined in a sensitivity analysis for $t = -n$ varying from $t = -1$ to $t = -168$.

The blending time weights are determined for $x \in [-n .. -1]$ as follows:

$$f_{linear}(x) = \frac{x + |n|}{\sum\limits_{i=n}^{-1} i + |n|} \tag{1}$$

$$f_{constant}(x) = \frac{1}{|n|} \tag{2}$$

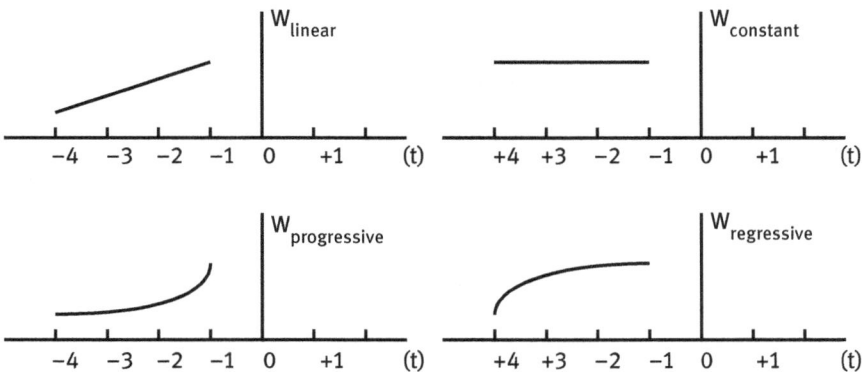

Fig. 5. Schematic visualization of linear (top left), constant (top right), progressive (bottom left) and regressive (bottom right) weighting.

$$f_{progressive}(x) = \frac{(x + |n|)^2}{\sum\limits_{i=n}^{-1} (i + |n|)^2} \tag{3}$$

$$f_{regressive}(x) = \frac{\sqrt{x + |n|}}{\sum\limits_{i=n}^{-1} \sqrt{i + |n|}} \tag{4}$$

It should be noted that $\sum\limits_{i=n}^{-1} f(x) = 1$ for $x \in [-n .. -1]$ with (1) to (4). The blending algorithm is applied with rolling horizon for a combined forecast.

3.5 Evaluation metrics

Commonly used metrics are employed for evaluating forecasts. The *mean average percentage error* (MAPE) and the *normalized rooted mean square error* (NRMSE) allow to compare the results to other case studies. The NRMSE is defined as

$$\text{NRMSE} = \sqrt{\frac{\sum\limits_{t=1}^{n} (\overline{y}_t - y_t)^2}{\sum\limits_{t=1}^{n} y_t^2}}$$

where y_t is the load profile from the ENTSO-E time series and $\overline{y}_t$ is the forecast load at time t. The MAPE is defined as

$$\text{MAPE} = \frac{1}{n} \sqrt{\sum\limits_{t=1}^{n} \left| \frac{\overline{y}_t - y_t}{y_t} \right|}$$

4 Results and discussion

Results of the prognoses for 2014, including national holidays, are shown in Tab. 2 for typical power exchange products such as base, peak and off-peak.[7] The column "week base" represents the best and the worst week (base product). It should be noted that the last week of the year had the worst forecasting performance. This week is characterized by national holidays, holiday season and bridging days making it

---

[7] Base product range from 0:00 h to 24:00 h every day. The peak time range is defined from 8:00 to 20:00 h on working days. Off-Peak is defined as 0:00 h to 8:00 h and 20:00 h to 24:00 h for all weekdays including weekends as well as national holidays.

Tab. 2. Results of model arrays (input) and blended forecasts.

2014		MAPE [%]			
Forecast	Model	base	peak	off-peak	weeks base
Input	REG	2.36	2.28	2.40	1.19–6.10
	TYPE	2.21	1.88	2.39	0.99–6.71
	ANN	2.20	1.87	2.38	0.85–6.71
	MVLR	2.13	1.94	2.24	1.04–6.69
	SVR	2.10	1.98	2.17	0.90–6.32
Blend	Blend 1 h	1.46	1.11	1.66	0.79–3.22
	Blend 8 h	1.63	1.39	1.77	0.90–5.87
	Blend 24 h	1.72	1.53	1.83	0.89–7.16

2014		NRMSE [$\cdot 10^{-2}$]			
Forecast	Model	base	peak	off-peak	weeks base
Input	REG	2.99	2.91	3.05	1.49–7.50
	TYPE	2.73	2.41	3.00	1.25–7.79
	ANN	2.73	2.41	2.99	1.06–7.78
	MVLR	2.74	2.59	2.87	1.40–8.53
	SVR	2.70	2.55	2.82	1.80–8.33
Blend	Blend 1 h	1.97	1.53	2.30	0.98–4.71
	Blend 8 h	2.17	1.94	2.37	1.08–7.62
	Blend 24 h	2.24	2.05	2.40	1.11–8.63

difficult to precisely forecast electrical demand. Furthermore, the data basis for model training and calibration is rather small. The input models are listed in increasing order (base MAPE): While the *REG-Model* has a MAPE of 2.36 % the *SVR-Model* reaches 2.10 %. The *SVR-Model* outperformed the other input models for base and off-peak forecasts. The best peak forecast comes from the *ANN-Model* with a MAPE of 1.87 %. The best and the worst weekly base forecast with a MAPE of 0.85 % and 6.71 % both generated by the *ANN-Model* is a detail of some interest.

The *Blend 1 h* forecast has a base MAPE of 1.46 % for the whole year and a weekly rating range from 0.79 % to 3.22 %. The day ahead *Blend 24 h* forecast also shows better results than the sole input models. The MAPE is 1.72 % for the year and the best week results in a MAPE of 0.89 %. Blending the available input forecasts increases the overall quality of the exemplary 1 h, 8 h intraday and the day ahead forecast.

Table 3 lists different results for blending models. The table also gives an indication for optimal analysis time ranges (ATR) for an ex-post data analysis. While in this example 1h intraday forecasts perform best by taking the last two hours progressively into consideration, the 8 h intraday forecast delivers best results by analyzing the last two days with a linear weighting strategy. The day ahead forecast, with an offset of 14 h (refer to section 3.4), performs best with a progressive weighting and 107 h ex-post ATR.

Tab. 3. Comparing different weighting strategies for 2014 base.

MAPE [%] | NRMSE [·10^{-2}]	lin		const.		prog.		reg.	
Blend 1 h; 2 h ATR	1.47	1.98	1.50	2.01	1.46	1.97	1.49	1.99
Blend 8 h; 47 h ATR	1.63	2.17	1.64	2.19	1.64	2.18	1.64	2.18
Blend 24 h; 107 h ATR	1.73	2.25	1.76	2.28	1.72	2.24	1.74	2.26

Table 4 shows the composition of the *Blend 1 h* and *Blend 24 h* forecast for the whole year. While the input forecast with the highest MAPE and NRMSE (*SVR-Model*) has the highest share (27 %) the other input models also contribute to the *Blend* prognoses.

Tab. 4. Composition of *Blend 1 h* and 24h from input forecasts.

Model	REG	TYPE	ANN	MVLR	SVR
1h [%]	17.11	21.85	12.44	22.52	26.57
24h [%]	16.73	21.26	12.86	21.97	27.66

Figure 6 gives a visual impression of how forecasting quality depends on ex-post Analysis Time Range. While forecasting quality suffers with a short ex-post ATR, the graphic shows that forecasting quality is enhanced by taking more ex-post information into the weighting consideration. The black grid represents the overall quality of the input *SVR-Model* as a comparison, underlining the strength of the introduced blending approach.

5 Conclusion

This paper's methodology is based on combining different forecasting methodologies, data decomposition to train model arrays for individual seasons and type days using fuzzy sets to compose forecasts from the model array, and a blending algorithm that learns from the past in order to assemble available input forecast for better intraday and day ahead forecasts.

Results represented verify the hypothesis form page 3. Blending of input forecasts, based on an ex-post analysis, increases overall accuracy compared to sole prognoses from model arrays. Input forecasts with mediocre results, such as the *REG-Model*, contribute their significant share to blended forecasts. That verifies the second hypothesis of this paper.

The results of this case study allow following conclusions: a) measuring the quality of forecasting methodologies with evaluation metrics may be misleading

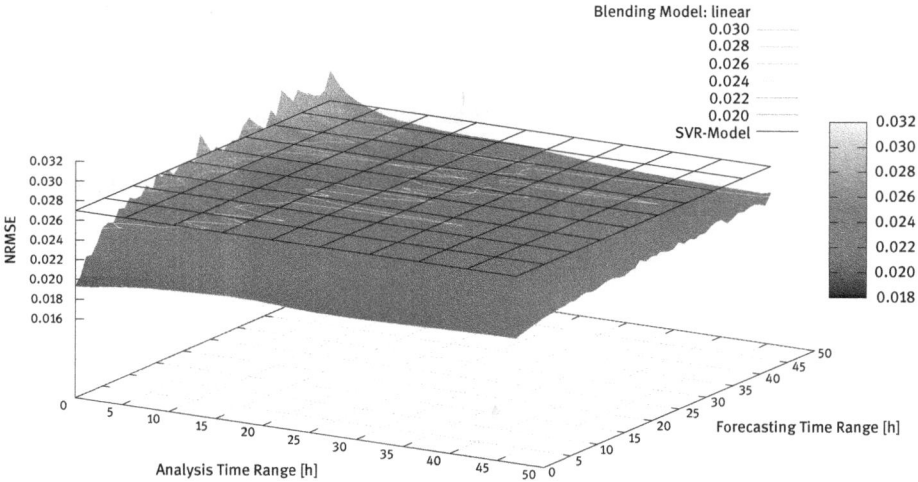

Fig. 6. Visualization of the forecasting quality depending on Analysis Time Range and Forecasting Time Range. The black grid represents the overall forecasting quality of the best input forecast (*SVR-Model*).

as different methodologies have individual strengths and weaknesses, b) ex-post analysis based blending combines the advantages of input forecasts and c) the most complex models do not prevail in all situations (comparing e. g. the weekly results range from the *REG-* and the *SVR-Model*).

6 Outlook

The blending approach allows to continuously update forecasts. Each new weighted forecast will result in position changes. These position deltas should be traded on intraday power markets such as the European EPEXSpot. As electrical demand is mainly inflexible and unbalanced areas are not an option, a continuously updated position will impose high work loads on intraday traders, especially on quarter hour products. With the overall increase of liquidity and volatility on power markets, as well as the introduction of new products, such as half and quarter hours, complexity and risks will increase.

An automatized process of intraday power trading is a logical step [31]. While automated / algorithmic trading is common in the financial sector [32], algorithmic trading is a new issue in the intraday power trading sector. Algorithmic trading is a domain for financial products. Risks are mainly money-related, while intraday power trading is a "physical" trade. Delivery of power is due 20 min (Nordpool Spot), 3o min

(EpexSpot) after gate closure. That's why reservations about introducing algorithmic trading in intraday power markets are still high.

"Fat Finger Trades" and resulting "Flash Crashes" may be a problem of stock markets, whereas European power utilities make security of supply their primary target. Therefore, it is mandatory to keep experienced intraday power traders in control of trading activities. One possible approach is shown in the following Fig. 7:

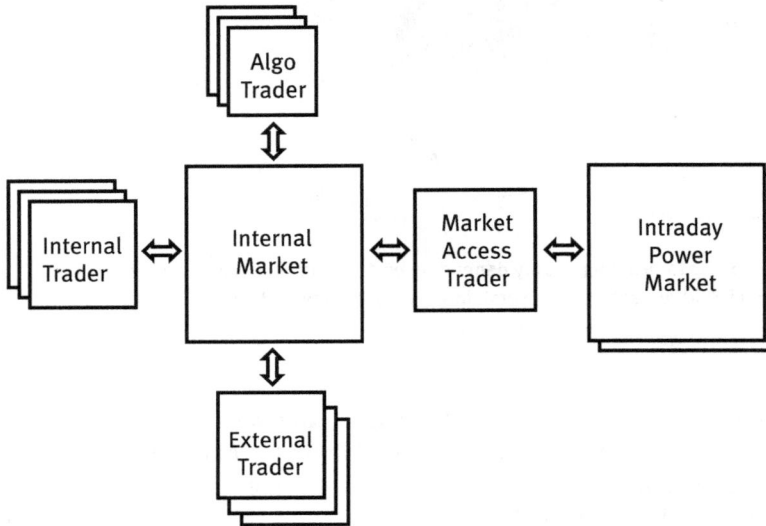

Fig. 7. Visualization of an Internal Market Power Trading regime.

The central component is the Internal Market (IM). The IM allows Internal Traders, supervising different portfolios of their clients (Trading as a Service), and External Traders (Market Access as a Service), usually smaller utilities without own direct market access, to trade. Algorithmic trading strategies have access to IM only. Market data is passed from external intraday power markets to the IM. Matching of ask / bid orders is granted on the IM, saving trading volumes and fees. Unmatched orders may be routed[8] to the external Intraday Power Market by authorized (licensed) Market Access Traders (MAT) only. Thus the MAT is in control of algorithmic trading activities, as he has access to the sandboxed trading environment IM.

Development of trading strategies for different use cases is a key aspect of algorithmic power trading. Before trading algorithms are released into the wild, they

---

8 It should be noted that a fully automatic routing can be configured.

need to undergo severe testing and quality evaluations. Two testing approaches follow:

1. Using the IM as a test environment by connecting external Intraday Power Markets as a read only price feed. Liquidity on the IM is generated by an agent algorithm placing fictional buy and sell orders with different prices and volumes. A second agent reduces liquidity by randomly accepting orders [33]. These two agents may be configured to replay market scenarios or place orders according to the real time data feed from the external market. Algorithmic trading strategies have to prove their efficiency and can be followed in real time by monitoring.

2. The second approach is a back testing environment. Long term order book recordings are replayed in the back testing environment in an accelerated manner. While Intraday Power Markets suffer from low liquidity (compared to stock markets) market impact of trading activities is an issue. Hence, trading actions by the algorithmic trader have to be considered.

While the IM may be used as a test environment for real time evaluations, the back testing environment offers possibilities to evaluate encoded trading strategies for longer time ranges. By varying settings of trading algorithms strategies beating the volume weighted average price (VWAP) may be identified.[9] However, there is a great risk that trading strategies are adapted to the testing environment, delivering high degrees of performance, but fail once they are used on real markets. This phenomena is called "over fitting". Besides over fitting, additional risks for algorithmic trading strategies are described in [34].

A common use case for algorithmic power trading on intraday markets is to close open positions caused by updated forecasts due to changes of feed in renewable sources or due to changes in the electrical load as described in this paper. These Position Trading Algorithms (PTA) attempt to close positions with hourly, half-hourly and quarter-hourly products advantageous to the VWAP. PTA may be constrained by limits which reduce their options to fully close positions on gate closure. However, PTA require models evaluating the market situation (risk model), forecast developments of the VWAP (alpha model), and an execution model [32].

Another common use case is the identification of arbitrage opportunities. Automated Trading Algorithms are able to identify cross market arbitrage opportunities by comparing EpexSpot and Nordpool Spot markets and cross product opportunities by comparing hourly products vs. quarter- hourly products. Of course Arbitrage algorithms need to include transaction costs in their evaluations.

While it would exceed the scope of this paper to go more into detail on algorithmic trading for intraday power markets, we plan to evaluate different PTA strategies on a sandboxed Internal Market for further publications.

---

9 In case of a position trading algorithm.

Acknowledgement: This paper is an extended version of [35] and [36]. We acknowledge the ENTSOE, DWD, Destatis for making the used data publicly available. Furthermore we thank Mr. U. Groth for proofreading and EXXETA for their support.

Bibliography

[1]   Commission. O.T. European Parliament. Regulation (EC) No 714/2009 of the European Parliament and of the Council of 13 July 2009 on conditions for access to the network for cross-border exchanges in electricity and repealing Regulation (EC) No 1228/2003, 2009.

[2]   B.F. W.U.E. BMWi. Ein Strommarkt für die Energiewende (Weissbuch), 2014.

[3]   Bundesnetzagentur. Monitoringbericht 2014, 2014.

[4]   F. Scheller, K. Keitsch, D. G. Reichelt, S. Dienst, S. Kuehne, H. Kondziella and T. Bruckner Evaluation von Geschäftsmodellen im liberalisierten Energiemarkt. *BWK*, 11:36–38, 2015.

[5]   H.K. Alfares and M. Nazeeruddin. Electric load forecasting: literature survey and classification of methods. *International Journal of Systems Science*, 33(1)1:23–34, 2002.

[6]   S. Khatoon and A.K. Singh, *et al.* Analysis and comparison of various methods available for load forecasting: An overview. *Computational Intelligence on Power, Energy and Controls with their impact on Humanity (CIPECH), 2014 Innovative Applications of*, 243–247, 2014.

[7]   P.K. Dash, A.C. Liew and S. Rahman, Fuzzy neural network and fuzzy expert system for load forecasting. *IEE Proceedings-Generation, Transmission and Distribution*, 143(1):106–114, 1996.

[8]   R. Barzamini, F. Hajati, S. Gheisari and M.B. Motamadine. Short term load forecasting using multi-layer perception and fuzzy inference systems for Islamic countries. *Journal of Applied Sciences*, 12:40–47, 2012.

[9]   M. Mordjaoui and B. Boudjema. Forecasting and modelling electricity demand using Anfis predictor. *Journal of Mathematics and Statistics*, 7(4): 275–281, 2011.

[10] M. Hell, P. Costa and F. Gomide. Participatory learning in the neurofuzzy short-term load forecasting. *Computational Intelligence for Engineering Solutions (CIES), 2014 IEEE Symposium on*, 176–182, 2014.

[11] S. Hemachandra and R.V.S. Satyanarayanab. Computational Intelligence for Prediction of Electrical Load, 387–391, 2014.

[12] K. Kampouropoulos, F. Andrade Rengifo, A. Garcia Espinosa, J.L. Romeral Martinez, *et al.* A combined methodology of adaptive neuro-fuzzy inference system and genetic algorithm for short-term energy forecasting. *Advances in Electrical and Computer Engineering*, 14:9–14, 2014.

[13] B. Islam, Z. Baharudin, Q. Raza and P. Nallagownden A Hybrid Neural Network and Genetic Algorithm Based Model for Short Term Load Forecast. *Research Journal of Applied Sciences, Engineering and Technology*, 13(7):2667–2673, 2014.

[14] V. Miryazdi, M. Ghasemzadeh, A.M. Latif, *et al.* Add a New Input to Neural Network with Genetic Learning Algorithm to Improve Short-Term Load Forecasting. *International Journal of Scientific Engineering and Technology*, 4(5):338–341, 2015.

[15] M.K. Abd. Electricity load forecasting based on framelet neural network technique. *American Journal of Applied Sciences*, 6(5):970–973, 2009.

[16] D.K. Chaturvedi, S.A. Premdayal, A. Chandiok, *et al.* Short-term load forecasting using soft computing techniques. *International Journal of Communications, Network and System Sciences*, 3(3):273–279, 2010.

[17] S. Li, P. Wang and L. Goel. A Novel Wavelet-Based Ensemble Method for Short-Term Load Forecasting with Hybrid Neural Networks and Feature Selection. *IEEE Transactions on power systems*, 3(31):1788–1798, 2015.

[18] N. Kumar Singh, A. Kumar Singh and P. Kumar. PSO optimized radial basis function neural network based electric load forecasting model. *Power Engineering Conference (AUPEC), 2014 Australasian Universities*, 1–6, 2014.

[19] X. Guo, X. Guo and J. Su. Improved Support Vector Machine Short-term Power Load Forecast Model Based on Particle Swarm Optimization Parameters. *Journal of Applied Sciences*, 13(9):1467–1472, 2013.

[20] K. Yang and S. Liu. A Hybrid Model for Short-Term Load Forecasting Based on Non-Parametric Error Correction. *International Journal of Multimedia and Ubiquitous Engineering*, 10(6):329–340, 2015.

[21] S. Fan and L. Chen. Short-term load forecasting based on an adaptive hybrid method. *IEEE Transactions on Power Systems*, 21(1):392–401, 2006.

[22] A. Jain and B. Satish. Clustering based short term load forecasting using support vector machines. *PowerTech, 2009 IEEE Bucharest*, 1–8, 2009.

[23] A. Khotanzad, R.-C. Hwang, A. Abaye and D. Maratukulam. An adaptive modular artificial neural network hourly load forecaster and its implementation at electric utilities. *IEEE Transactions on Power Systems*, 10(3):1716–1722, 1995.

[24] M. Devaine, P. Gaillard, Y. Goude and G. Stoltz. Forecasting electricity consumption by aggregating specialized experts. *Machine Learning*, 90(2): 231–260, 2013.

[25] E.N. of Transmission System Operators for Electricity (ENTSO-E). Detailed Data Descriptions, tech. rep., 02 2014.

[26] M. Schumacher and L. Hirth. How much Electricity do we Consume? A Guide to German and European Electricity Consumption and Generation Data. *Climate Change and Sustainable Development*, Series Editor: Carlo Carraro, 88:1–33, 2015.

[27] S.G.V. (Herausgeber). Produktionsindex. *Gabler Wirtschaftslexikon*, 2015.

[28] M.H. Choueiki, C.A. Mount-Campbell and S.C. Ahalt. Building aquasi optimal neural network to solve the short-term load forecasting problem. *IEEE Transactions on Power Systems*, 12(4):1432–1439, 1997.

[29] C.-C. Chang and C.-J. Lin. LIBSVM: A library for support vector machines. *ACM Transactions on Intelligent Systems and Technology*, 2(3):1–27, 2011.

[30] C.-W. Hsu, C.-C. Chang, C.-J. Lin. *et al.* A practical guide to support vector classification. 2003.

[31] K. Keitsch, N. Bornhöft, J. Becker and A. Wielend. Algorithmic Trading - Der Einsatz von Handelsalgorithmen in der Energiewirtschaft. *emw - Energie. Markt. Wettbewerb.*, 2:48–51, 2017.

[32] P. Treleaven, M. Galas and V. Lalchand. Algorithmic trading review. *Communications of the ACM*, 56(11):76–85, 2013.

[33] R. Hu and S.M. Watt. An agent-based financial market simulator for evaluation of algorithmic trading strategies. *6th International Conference on Advances in System Simulation*, 221–227, 2014.

[34] U.J. Hammerich Beyond backtesting: The historical evidence trap. *Available at SSRN 2607237*, 2015.

[35] K. Keitsch and T. Bruckner. Modular Electrical Demand Forecasting Framework - A Novel Hybrid Model Approach. *13th International Multi-Conference on Systems, Signals and Devices (SSD 2016)*, (Leipzig, Germany), 2016.

[36] K. Keitsch and T. Bruckner. SAWing on short term load forecasting errors: Increasing the accuracy with self adaptive weighting. *IEEE PES Innovative Smart Grid Technologies - Asia*, (Melbourne, Australia), 1008–1013, 2016.

Biographies

Krischan Keitsch is working at the EXXETA AG as senior consultant in the department for Energy Trading and Risk Management. Prior he has been working at the Fraunhofer IMW Center for International Management and Knowledge Economy (Leipzig, Germany) as a research associate and doctoral candidate in the business unit "Energy Management and Energy Economics" since July 2014. Prior to that, he worked in various positions over a period of about ten years: as consultant at GOagile! AG, as project manager in "energy management & market environment" at EnBW Holding AG in the area of financial management, and as project manager and power trader in the field of "energy planning and power trading" at EnBW Trading GmbH. He studied Energy and Process Engineering at Technische Universität Berlin. His first work experience was gained as a member of a project team in the field of "energy efficiency in the electricity sector" at Deutsche Energie-Agentur GmbH (German Energy Agency – dena). Krischan A. Keitsch was born in Berlin, Germany, February the 6th, 1976.

Prof. Dr. Thomas Bruckner is a physicist by training who has written his PhD thesis in theoretical physics about "Dynamic Energy and Cost Optimization of Regional Energy Supply Systems" in 1998. Between 1996 and 2000 he was a researcher at the Potsdam Institute for Climate Impact Research, where he was involved in an international project for the integrated assessment of global climate protection strategies. Afterwards, he led the research group "Energy System Optimization and Climate Protection" at the Institute for Energy Technology at the Technical University of Berlin. Since 2008, he is the Chair for Energy Management and Sustainability at the Faculty of Economics and Business Management, Leipzig University, acting director of the Institute for Infrastructure and Resources Management (since 2009) and leader of several interdisciplinary projects within the field of energy system and integrated assessment modeling. As a member of the Intergovernmental Panel on Climate Change (IPCC, since 2008) he was coordinating lead author of the chapter on "Energy Systems" in the 5th IPCC assessment report (in print) and one of the lead authors of the IPCC Special Report on Renewable Energies published in 2011.

W. Khemiri, A. Sakly and M.F. Mimouni

Comparison of three Loss Optimization Techniques of FOC Induction Motor Drive

Abstract: This paper presents a comparative study of an analytical method and particle swarm optimization method for minimum-energy loss for a field oriented control induction motor drive in transient regime. Those suggested strategies are based on the optimization of the input active power to determine the optimal flux which minimizes the induction motor efficiency. Comparing with the conventional field oriented control it has been observed that the Particle Swarm Optimization based algorithm is much more efficient than the other methods.

Keywords: Induction motor, dynamic regime, field oriented control, particle swarm optimization, energy optimization.

1 Introduction

Three phase induction motor (IM) are robust machines considered as the universal work horses of industry. Hence, they cover up heavy industrial applications [1].

However, as majority of electrical energy generated is consumed by motor drives, so slight improvement in their efficiency can save a good amount of energy [2]. Recently, on account of the limitation of energy sources, the industry evokes one of the major serious problems, which is the energy consumption. Several strategies have been suggested in order to cure this consumption which does not cease growing. Among them, we are interested in those allowing the maximization of the efficiency, through the power-losses minimization in the machine named loss model control technique (LMC). This can be achieved very efficiently for induction motor using Field Oriented Control (FOC). LMC techniques can be classified into two categories namely conventional and numerical. In the conventional LMC techniques, based on the motor loss model, the FOC generates the required reference currents based on the reference torque. Conventional LMC techniques are available in the literature. Their common characteristic is their dependence on IM parameters and specific loss equations. In [3], an LMC algorithm is developed to optimize the efficiency of a FOC IM drive. A loss model derived by considering the stator and rotor copper losses of the motor. To minimize these losses an expression of the optimal flux is determined.

W. Khemiri, A. Sakly and M.F. Mimouni: National Engineering school of Monastir (ENIM), Research Unit: ESIER Monastir, Tunisia, W. Khemiri, email: khemiriwahiba@yahoo.fr, A. Sakly, email: sakly_anis@yahoo.fr, M.F. Mimouni, email: Mfaouzi.mimouni@enim.rnu.tn

De Gruyter Oldenbourg, ASSD – Advances in Systems, Signals and Devices, Volume 7, 2018, pp. 19–34.
https://doi.org/10.1515/9783110470529-002

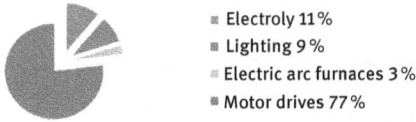

■ Electroly 11%
■ Lighting 9%
▪ Electric arc furnaces 3%
■ Motor drives 77%

Fig. 1. IM industrial applications.

Similar approaches are applied in [4]–[7]. Many recent developments in science and engineering demand numerical techniques for IM efficiency optimization. For that purpose, many artificial intelligence have been applied in literature such as: artificial neural networks (ANN) [8], fuzzy logic (FL) [9], and genetic algorithm (GA) [10]. Now a day, a new evolutionary computing technique called particle swarm optimization (PSO), has been proposed and introduced [11]. PSO approach is proposed to solve the problem of the optimization of the IM energy in the dynamic regime [12]. The main objective of this work is to compare efficiency of both PSO and the analytical optimization technique [13] applied to a FOC IM drive. Therefore, these two methods have been applied to search for the optimal flux of FOC IM. The two approaches are tested on a 1.5 kw IM and the simulation results obtained are compared with the conventional FOC method. Thus, this paper is organized in three sections. In the first section, an overview of the three suggested strategies, namely conventional FOC, Optimized FOC (OFOC) and PSO based FOC, is presented. The second section is devoted to the problem formulation with variable and constraints. In the third section, we present a comparative study between these three strategies through simulation results.

N.B.: All citations and labels should start with the name or the surname of the first author. See the current tex-file.

2 Problem formulation

The main objective is to minimize the energy consumption of an IM. The proposed methods enable one to optimize the energy consumption of the IM in terms of a cost function while satisfying two constraints. Mathematically, based on the reduced dynamic model of the IM, the problem can be formulated by developing the IM energetic model.

2.1 Reduced dynamic model of the IM

Based on the usual assumptions for the Blondel-Park's transformation [14] and the "singular perturbation" theorem [15], the reduced model of the IM for a load torque which is proportional to the motor speed [16] is presented as follows:

$$\dot{\Phi}_r^{(d,q)} = -(aI + \omega_s J)\Phi_r^{(d,q)} + bU \tag{1}$$

$$\dot{\Omega} = -\frac{K_l}{J_m}\Omega + \frac{cU^T J \Omega_r^{(d,q)}}{J_m} \tag{2}$$

where:

$$U = \begin{bmatrix} u_1 \\ u_2 \end{bmatrix} = \begin{bmatrix} I_{sd} \\ I_{sq} \end{bmatrix}, I_s^{d,q} = \begin{bmatrix} I_{sd} \\ I_{sq} \end{bmatrix}, \Phi_r^{d,q} = \begin{bmatrix} \Phi_{rd} \\ \Phi_{rq} \end{bmatrix}, V_s^{d,q} = \begin{bmatrix} V_{sd} \\ \Phi_{sq} \end{bmatrix} \sqcap$$

$$I = \begin{bmatrix} 1 & 0 \\ 0 & 1 \end{bmatrix}, J = \begin{bmatrix} 0 & -1 \\ 1 & 0 \end{bmatrix}$$

$$\rho = \arctan\frac{\Phi_{ra}}{\Phi_{rb}}, c = \frac{3}{2}p\frac{M}{L_r}, b = aM\gamma = \frac{1}{\sigma L_s}\left(R_s + \frac{M^2}{L_r^2}R_r\right), \eta = \frac{M}{\sigma L_s L_r}, a = \frac{R_r}{L_r}$$

with: R_r, R_s : rotor and stator resistance, L_r, L_s: rotor and stator inductance, M: magnetizing inductance, ω_e: angular supply frequency, σ: dispersion coefficient, ω_s, $\dot{\rho}$: angular, sleep frequency, Ω, ω_r: motor speed, p: number of poles, K_e,K_h: eddy and hysteresis current coefficients, I_{sd}, I_{sq}: magnetizing and torque currents in d-q axis, I_{rd}, I_{rq}: rotor currents in d-q axis, V_{sd}, V_{sq}: stator voltages in d-q axis, Φ_r, Φ_m: rotor and air gap fluxes, Φ_{rd}, Φ_{rq}: rotor fluxes in d-q axis, y: electromagnetic torque, I: 2×2 unit matrix, J: 2×2 skew symmetric matrix, P_a: input active power, w: Stored magnetic energy, P_J : Copper losses and P_m: Mechanical power, J_m: is the total moment of inertia of the rotor and fly-wheel masses reduced to the motor shaft and K_l: is the load torque coefficient.

2.2 Energetic model

The instantaneous input active power of an IM expressed in the rotating dq-frame is given by [13]:

$$P_a = \frac{dW}{dt} + P_J + P_m \tag{3}$$

when: The expression of copper losses is given by:

$$P_J = \frac{3}{2}\left(\left[R_s + Rr\left(\frac{M}{L_r}\right)^2\right](U_1^2 + U_2^2) + \frac{R_r}{L_r^2}\Phi_r^2 - \frac{1}{L_r}\frac{d\Phi_r^2}{dt}\right) \tag{4}$$

The mechanical power transmitted to the motor rotor, can be derived as:

$$P_m = \frac{3}{2}\left(\frac{M}{L_r}\right)\Phi_r U_2 \omega \tag{5}$$

The derivate of the stored magnetic can be expressed as follows:

$$\frac{dW}{dt} = \frac{3}{2}\left[\left(\frac{\sigma L_s}{2}\right)\left(U_1^2 + U_2^2\right) + \frac{R_r}{2L_r}\Phi_r^2\right] \tag{6}$$

2.3 Cost function design

The cost function represented by the integral of an index , it is given as follows:

$$J_r = \int_0^T f\left(I_{sd}, I_{sq}, \Phi_r, \Omega\right)dt \tag{7}$$

The considered cost function must be minimizing in transient state closed loop regime under the following dynamic constraint:

$$\dot{\Phi}_r = -a\Phi_r + bu_1 \tag{8}$$

$$\dot{\Omega} = -\frac{K_l}{J}\Omega + \frac{c}{J}u_2\Phi_r \tag{9}$$

The optimization problem stated in (7), (8) and (9) consists in seeking for a control optimal value U^*, defined for all $\Phi_r > 0$, as:

$$u_1^* = \frac{1}{b}\left(\dot{\Phi}_r^* + a\Phi_r^*\right) \tag{10}$$

$$u_2^* = \frac{J_m}{c\Phi_r^*}\left(\dot{\Omega}^* + \frac{K_l}{J_m}\Omega^*\right) \tag{11}$$

Therefore, the cost function is a function of only the two variables Φ_r and Ω so the optimizing problem aims to find the optimal steady state solution (Φ_r^*, Ω^*) which minimizes the cost function at the rated torque.

3 Overview of Some Loss Optimization Techniques

3.1 Conventional FOC

The FOC technique was introduced by Blaschke in 1972 [17]. By This type of control, the transient response of IM improves since IM can be controlled like DC machine where its torque component and field flux component are separated virtually and independent control of each component is possible [18]. FOC is based on phase transformation, where a three phase time and space variant system is transformed to a synchronously rotating, time invariant system, leading to a structure similar to that of

DC machine [19]. Thus, the FOC uses the dynamic mathematical model of the IM and decouples control of the flux and torque which makes the IM deliver excellent dynamic performance [20]. / The input of the FOC module is the reference speed; as shown in Fig. 2. The FOC generates the required reference currents to drive the motor. These currents are based on the flux level, which determines the values of direct current, and the reference torque which determines the values of quadrature current. The reference torque is calculated according to the speed error.

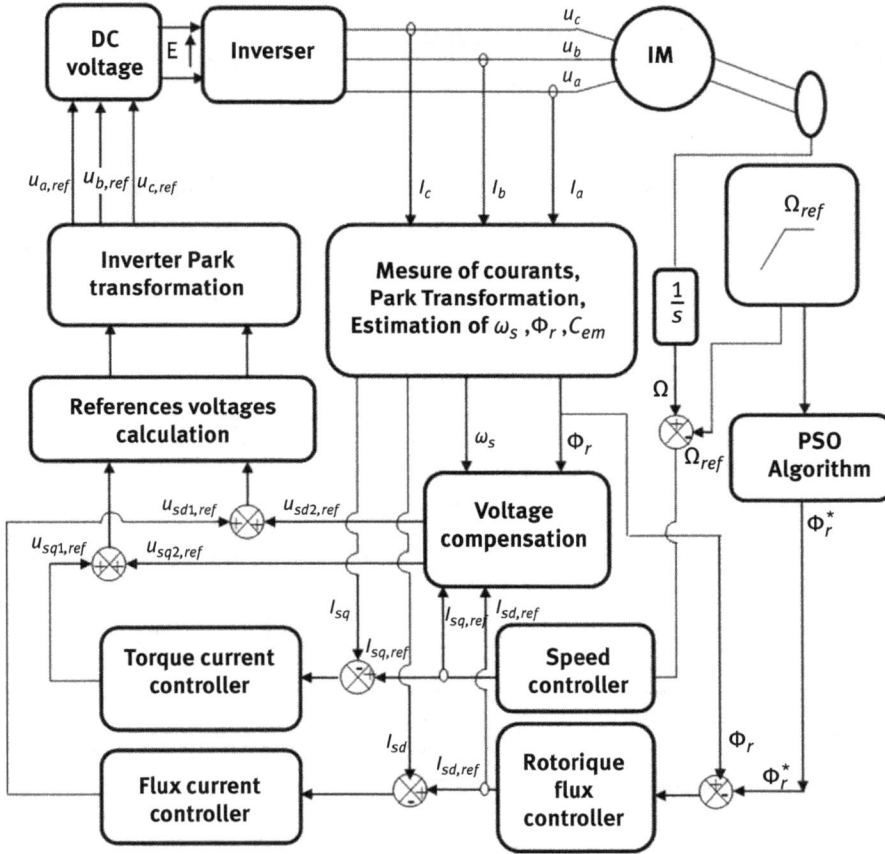

Fig. 2. FOC scheme.

The optimal OFOC, developed in [13], provides an analytical solution for expressing the trajectory of the rotor flux level reference according to the charge level and for a given IM speed profile. This method is based on solving the objective function defined in (7) using Euler-Lagrange equation. Solving this equation and the study of its properties are linked to the trajectory of speed. To speed trajectory are known in advance, we proposed an analytical solution for optimal trajectory of the rotor flux that minimizes the energy consumption of the asynchronous machine in transient regime. This solution, given by (12), is calculated by applying the method of integration by part during a fixed time interval. In order to implement these optimal trajectories in vector control of induction motor, we proposed an improvement in the pursuit of optimal reference flux via the use of a Deadbeat controller.

$\Phi_r(t) =$

$$\sqrt[6]{\frac{30}{a_2}\left[a_2\Phi_r^6(0)+\left(-\dot{\Phi}_r(0)\Phi_r^3(0)+\Phi_r^3(0)+\frac{1}{5}\Phi_r^5(0)\right)t+\left(\frac{\gamma_1 t^4}{12}+\frac{\gamma_2 t^3}{6}+\frac{\gamma_3 t^2}{2}\right)-k\right]} \quad (12)$$

where k is an appropriate positive constant.

The OFOC drive of the IM, given in Fig. 3, is initialized through applying a motor speed reference.

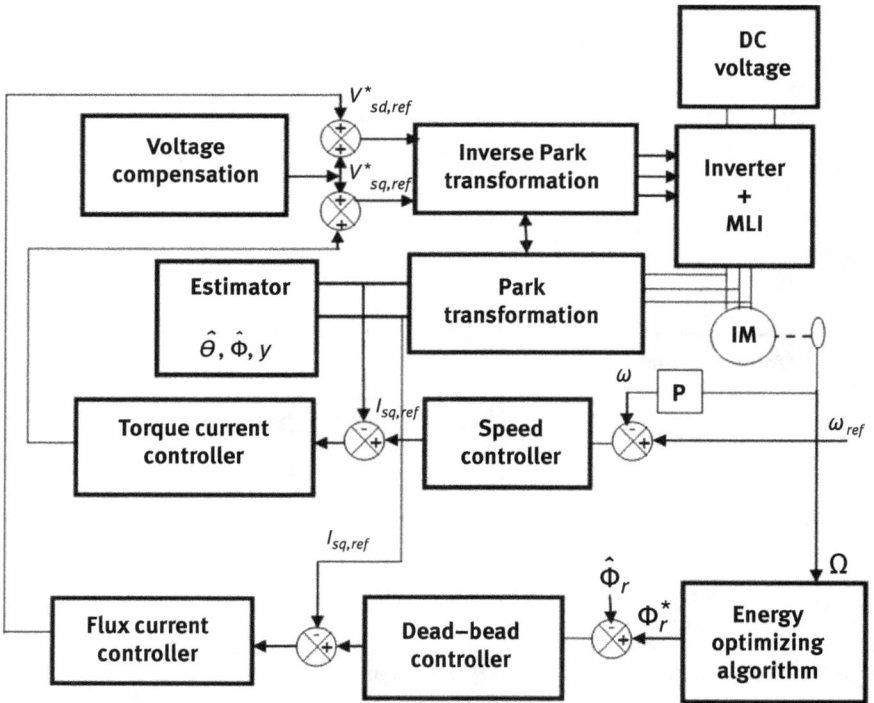

Fig. 3. Control system configuration.

An optimal rotor flux current by means of the deadbeat controller is delivered to the remaining part of FOC drive. On the other hand, a transient torque current reference will be delivered to the rest of the FOC drive.

3.2 PSO-based FOC

PSO is a new optimization method and was developed by Kennedy and Elberhart in 1995 [21] to imitate the motion of a flock of birds or insects. The birds are called as "particle" and constitute a swarm. PSO is initialized with a group of random particles (solutions) and then searches for optima by updating generations [22]. Each individual particle in PSO flies in the search space with a velocity which is adjusted dynamically according to its own flying experience and also its neighbor's flying experience [23]. In the PSO algorithm, the population has n particles and each particle is an m dimensional vector, where m is the number of the optimized parameters.

The process of the proposed algorithm is shown in Fig. 4 and can be described in the following steps:

Step 1: Initialization
- Initialize iteration counter $k = 0$.
- Initialize N random position of particles ($X_i^k; i = 1, 2, ..N$) and store them in X.
- Initialize N random velocities ($V_i^k; i = 1, 2, ..N$) and store them in V.
- Initialize N $Pbest$ ($Pbest_i^k; i = 1, 2, ..N$) and store them in $Pbest$.

Step 2: Velocity updating
To compute the velocity of each particle in the next stage we have to apply the equation (12).

Step 3: Position updating
Based on the updated velocities, the position of each particle is updated by the equation (13).

Step 4: Individual best and global best updating
Each particle is evaluated according to its objective function values at the position of particle i, then update the $Pbest$ as follow:

$$Pbest_i^{k+1} = X_i^{k+1}, if J_i^{k+1} < J_i^k$$

$Gbest$ at iteration $k + 1$ is set as the best evaluated position among $Pbest_i$

Step 5: Stopping criteria
The proposed PSO algorithm is stopped if the iteration reaches predefined maximum iteration kmax.

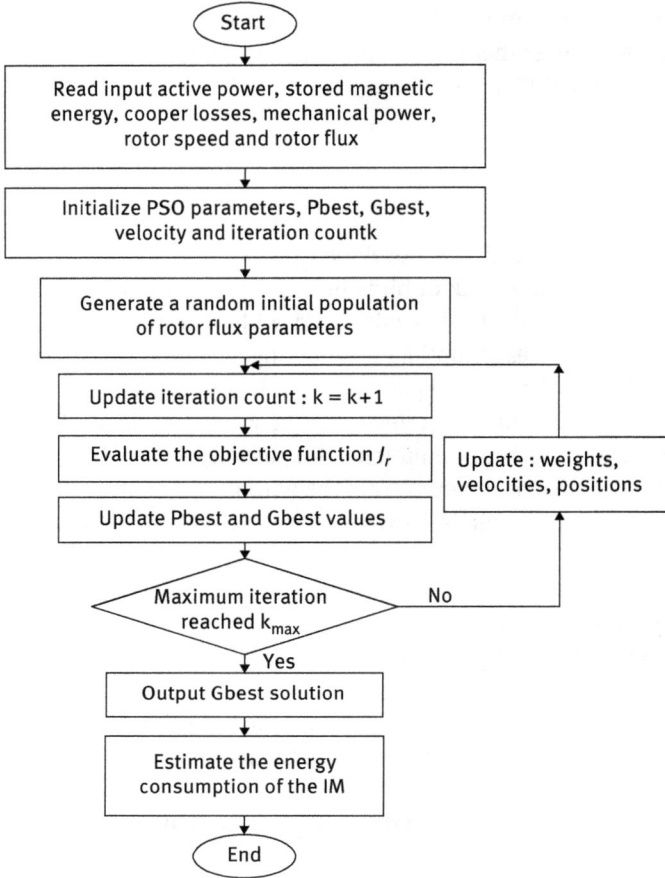

Fig. 4. Flowchart of the PSO algorithm.

The velocity and position of each particle are given by (13) and (14) respectively:

$$V_i^{k+1} = V_i^k + C_1 \times rand_1 \times \left(Pbest_i^k - X_i^k\right) + C_2 \times rand_2 \times \left(Gbest^k - X_i^k\right) \qquad (13)$$

$$X_i^{k+1} = X_i^k + V_i^k \qquad (14)$$

where: X_i^k is the current position of the particle i at iteration k, *Pbest* denotes the best previous position of the particle, *Gbest* represents the index of the best particle among all the particles, C_1 and C_2 present the weighted of stochastic acceleration terms that permit each particle toward the *Pbest$_i$* and the *Gbest* position. *rand$_j$* is a random number between 0 and 1. The inertia weight factor W is considered to decrease linearly [24] from its final value ($Wmax$) to the initial one ($Wmin$) as follows:

$$W^k = W_{max} - \frac{W_{max} - W_{min}}{k_{max}} \times k \tag{15}$$

The process of the proposed FOC based on PSO algorithm [13] is shown in Fig. 5.

3.3 Comparison between the three above methods

In order to carry out a comparative analysis of FOC, OFOC and PSO based FOC; the behavior of the three methods has been studied and illustrated in table 3. Indeed the analysis of characteristics of each of these methods allows us to make out the following advantages and drawbacks:

-   The FOC has proven itself in terms of robustness and reliability although it is sensible to the IM parameters variation.

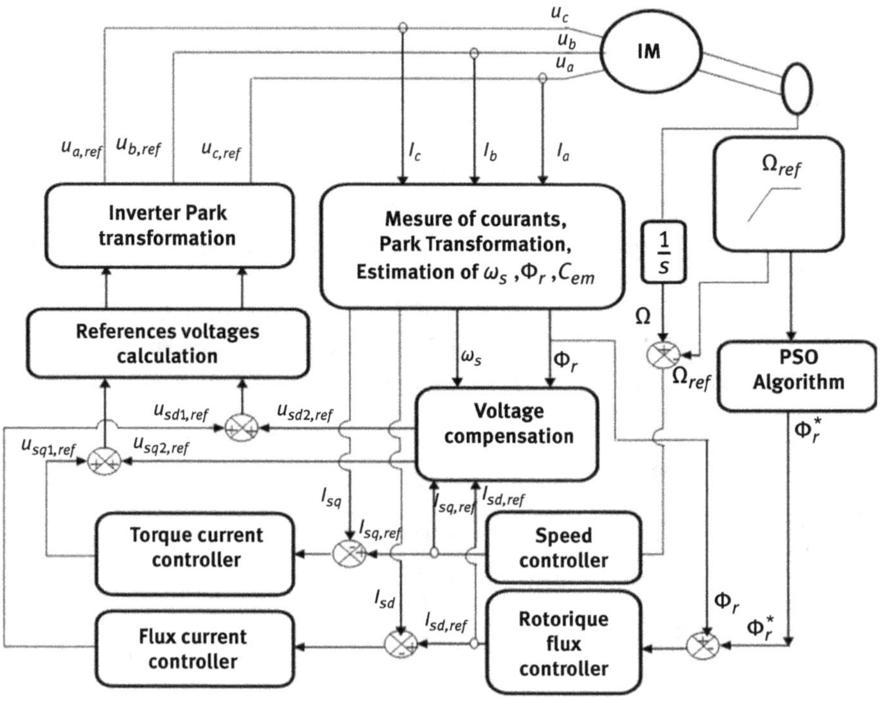

Fig. 5. Flowchart of the PSO algorithm.

Tab. 1. Summary of the comparison between the three methods.

Method features	FOC	OFOC	PSO
Intelligence	No	No	Yes
Dynamic response	Quick	Quick	Quick
Complexity of implementation	Easy	Easy	Very easy
Exactitude	Yes	Yes	No
Calculation	Simple	Complex	Very simple

- The PSO algorithm is characterized as simple in concept, easy to implement, flexible and computationally efficient [21], [24] and [25] despite of its stochastic nature.
- The analytical methods of the optimization problem have the advantage that their solutions can be used directly in real time, because it is an equation that gives the optimal trajectory of the reference flux. But it needs hard mathematical calculation.

It can be seen how the PSO is considerably superior to any traditional methods.

4 Simulation results

Some simulation tests have been designed and executed to obtain comparative results of the three methods. These results are approved on a three-phase IM, characterized by $220V/380V$, $50\,Hz$ and 4 poles, $1.5\,kW$. The motor parameters are presented in Table 2 and Table 3.

The comparison between the variations of the energy consumed by the IM using the PSO based approach to that using the OFOC and the conventional FOC, shows that a considerable energy and cost saving are achieved. Thus the efficiency that increases

Tab. 2. IM Simulation parameters, part 1.

Parameters	Values
Rotor resistance R_r	$4.2\,\Omega$
Stator resistance R_r	$6.06\,\Omega$
Rotor and Stator inductance L_r, L_s	$0.462\,\text{mH}$
Magnetizing inductance M	$0.44\,\text{H}$
Eddy and Hysteresis current coefficients K_e, K_h	0.0012
Total moment of inertia	$0.049\,\text{kgm}^2$

Tab. 3. IM Simulation parameters, part 3.

Parameters	Values
Population size	30
Particle dimension	4
Maximum iteration	10
C_1	1.5
C_2	1.5
W_{max}	0.9
W_{min}	0.4

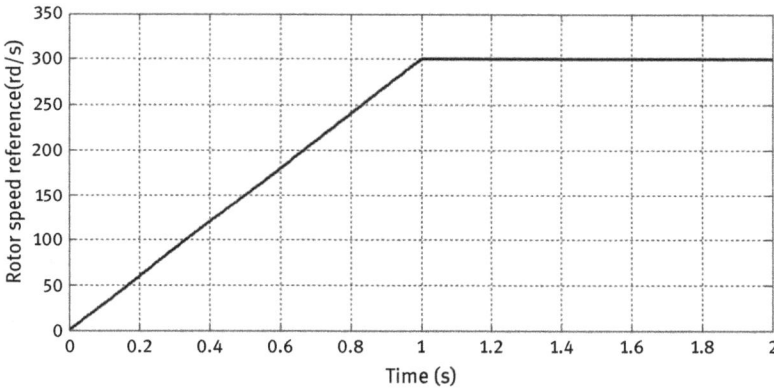

Fig. 6. Rotor speed reference.

with decreasing losses can be improved. Table 4 shows the total energy consumed by the IM comparison using few example of operating points.

In order to compare the efficiency of the three presented strategies, it is clear from Fig. 7 that a considerable energy saving is achieved in the case of using the PSO-based approach and the OFOC in comparison with the conventional FOC. From Fig. 8, 9 and 10 it can be seen that the IM energy consumption decrease is depended

Tab. 4. Comparison of the total energy consumed by the IM for different torque values.

Rotor speed (rd/s)	135	162	189	216	243	270	297	300
Energy saving PSO-FOC - FOC (J)	17,74	19,14	20,93	23,02	14,05	15,33	11,78	4,25
Energy saving OFOC - FOC (J)	2.92	3.87	6.63	14.9	9.86	23.86	53.03	18.98
Energy saving PSO-FOC - OFOC(J)	14,82	15,27	14,30	8,12	4,19	-8,53	-17,39	-14,73

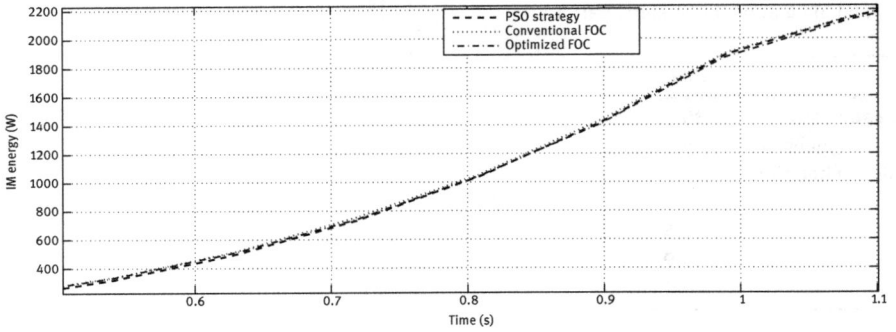

Fig. 7. Total energy consumed by the IM OFOC dive.

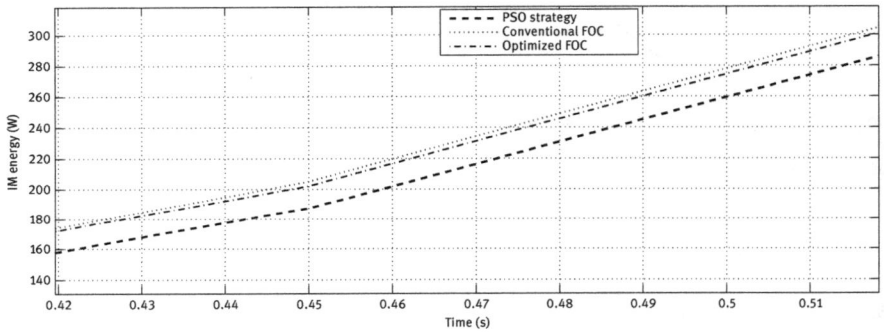

Fig. 8. Zooming of the total energy consumed by the IM OFOC dive for small torque values.

on the torque values. In fact, for small torque values (Fig. 8) we can see that the OFOC gives close results to the conventional FOC but when the torque values grow increase (Fig. 9) the PSO based-FOC is proved to give improved minimization of the IM energy consumption and when the torque reach its nominal value (Fig. 10) the OFOC becomes more efficient than the PSO based-FOC. These results are available in table 1. In fact, the PSO-based algorithm can reduce 8% of the total energy consumed by the IM compared with that of the conventional FOC and can reduce 6.9% compared with that of the optimized FOC.

5 Conclusion

The aim of this paper was to give a fair comparison between FOC, OFOC and PSO based FOC techniques. The synthesis of the simulation study reveals a slight advantage of

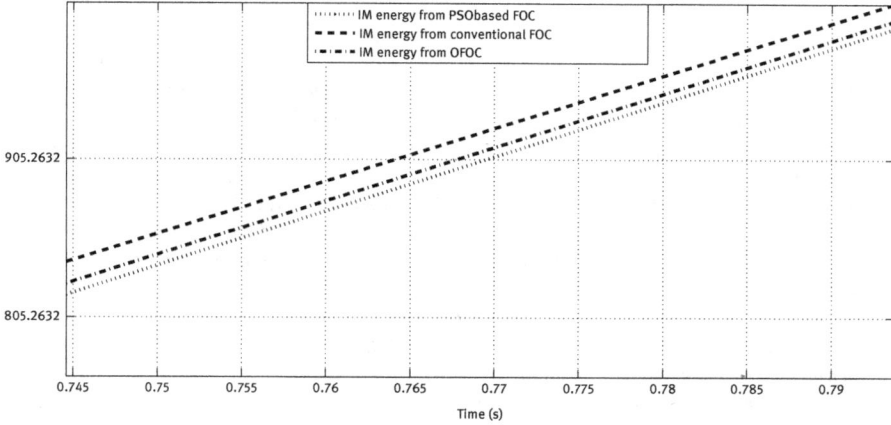

Fig. 9. Zooming of the total energy consumed by the IM OFOC dive for medium torque values.

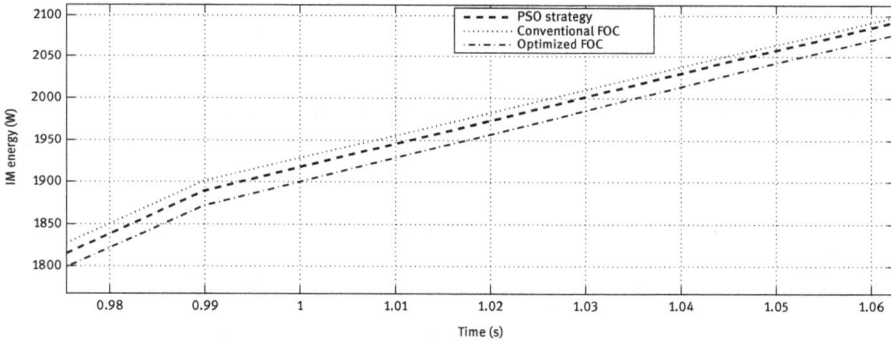

Fig. 10. Zooming of the total energy consumed by the IM OFOC dive for nominal torque values.

the PSO scheme compared to the OFOC scheme regarding the dynamic flux control performance. The description of both control schemes and their principle of operation has been presented. Several numerical simulations have been carried out in transient operating conditions. Summarizing, it can be said that both methods provide a high performance optimal rotor flux. The results confirm the potential of the proposed PSO-based FOC approach and reveal that it can reduce the motor energy consumption with more that 8 %, compared to the conventional FOC, which show its superiority over the other strategies for small and medium torque values. However, for nominal torque values the OFOC approach gives best results.

Bibliography

[1]   R. Vijayarkumar, M.P. MohanDass and S.A. Sreeja. An overview on Performance Improvement of an Induction Motors (IM)- A Review. *Int. Journal of Innovative Research in Electrical, Electronics, Instrumentation and Control Engineering*, 2:2019–2026, 2014.

[2]   B. Kumar, Y.K. Chauhan and V. Shrivastava. Assessment of a fuzzy logic based MRAS observer used in a photovoltaic array supplied AC drive. *Frontiers in Energy*, 8:81–89, 2014.

[3]   B. Kumar, Y.K. Chauhan and V. Shrivastava. Performance Analysis of Induction Motor Drive with Optimal Rotor Flux F or Energy Efficient Operation. *IEEE Int. Conference on Advanced Communication Control and Computing Technologies*, 319–322, 2014.

[4]   S. Thomas and R.A. Koshy. Efficiency Optimization with Improved Transient Performance of Indirect Vector Controlled Induction Motor Drives. *Int. Journal of Advanced Research in Electrical, Electronics and Instrumentation Engineering*, 2:374–385, 2013.

[5]   J-F. Stumper, A. Dotlinger and R. Kennel. Loss Minimization of Induction Machines in Dynamic Operation. *IEEE Trans. on Energy Conversion*, 28:726–735, 2013.

[6]   B. Blanuša and B. Knežević. Optimal Flux Control of Elevator Drive. *Int. Conf. on Information, Communication and Automation Technologies*, :1–6, 2013.

[7]   A.S. Chandrakanth, N. Kumar, T.R. Chelliah and S.P. Srivastava. Sensitivity Analysis of Model-Based Controller Applied to Loss Minimization of Induction Motor. *IEEE Int. Conf. on Power Electronics*, :1–5, 2012.

[8]   A. Taheri and H. Al-Jallad. Induction Motor Efficient Optimization Control Based on Neural Network. *Int. Journal on Technical and Physical Problems of Engineering*, 4:140–144, 2012.

[9]   Z. Rouabah, F. Zidani and B. Abdelhadi. Fuzzy Efficiency Enhancement of Induction Motor Drive. *Int. Conf. on Power Engineering, Energy and Electrical Drives*, 175–180, 2013.

[10] A. Dey, B. Singh, B. Dwived and D. Chandra. Vector Controlled Induction Motor Using Genetic Algorithm Tuned PI Speed Controller. *Electrical Power Quality and Utilisation Journal*, 17:3–8, 2009.

[11] J. Kennedy. The Particle Swarm: Social Adaptation of Knowledge. *IEEE Conf Evol. Compt. ICEC'97*, Indianapolis USA, 303–8, 1997.

[12] W. Khemiri, A. Sakly and M.F. Mimouni. Optimal Energy Consumption of an Induction Motor Using Particle Swarm Optimization. In *1st Int. Conf. on Information and Communication Technologies Innovation and Application (ICTIA)*, Sousse, Tunisia, 2014.

[13] W. Khemiri, A. Sakly and M.F. Mimouni. Minimun-Energy Consumption of an Induction Motor Operating in Dynamic Regime. 10th *Int. Multi-Conf. on System, Signal & Devices (SSD)*, Hammamet Tunisia, 2013.

[14] John Chiasson. Modelling and High-Performance Control of Electric Machines. *IEEE Press Series on Power Engineering, John Willey & sons, Inc.*, Hobokon, New Jersey, 2005.

[15] M.F. Mimouni and R. Dhifaoui. Modelling and Simulation of Double-Star Induction Machine Vector Control using Copper Losses Minimization and Parameters Estimation. *Int. Journal of Adaptative Control and Signal Processing*, :1–24, 2002.

[16] H.K. Khalil. Nonlinear Systems. *Mc Millan Pub*,New Jersey, 1996.

[17] F. Blaschke. The principle of field orientation as applied to the new TRANSVECTOR closed loop control system for rotating field machines. *Siemens Rev.*, 34:217–220, 1972.

[18] T. Banerjee, S. Choudhuri and J.N. Bera. Off-line Optimization of PI and PID Controller for a Vector Controlled Induction Motor Drive Using Genetic Algorithm. *Int. Conf. on Electrical Power and Energy Systems (ICPES)*, :192–195, August, 2010.

[19] T. Banerjee, S. Choudhuri, J.N. Bera and A. Maity. Off-line Optimization of PI and PID Controller for a Vector Controlled Induction Motor Drive Using PSO. In *Int. Conf. on Electrical and Computer Engineering (ICECE)*, :74–77, Dec 2010.

[20] C. Lai, K. Peng and G. Cao. Vector Control of Induction Motor based on Online Identification and Ant Colony Optimization. 2nd *Int. Conf. on Industrial and Information Systems (IIS)*, :206–209, 2010.

[21] J. Kennedy and R.C. Elberhart. Particle Swarm Optimization. *EEE Conf. on neural netwoks ICNN'95*, Perth Australia, pp1942–1948, 1995

[22] T. Banerjiee, S. Chowdhuri, G. Sarkar and J. Bera. Performance comparison between GA and PSO for optimization of PI and PID controller of direct FOC induction motor drive. *Int. J. of Scientific and Research Publications*, 2, Jul 2012.

[23] A. Mahesh and B. Singh. Vector control of induction motor using ANN and particle swarm optimization. *Int. J. of Engineering Technology and advanced Engineering*, 2:480–485, 2012.

[24] K.T. Chaturvedi, M. Pandit and I. Srivasatava. Particle Swarm Optimization with Time Varying Acceleration Coefficients from Non Convex Economic Power dispatch. *Electric Power Energy System*, :249–57, 2009.

[25] Y. Shi and R.C. Elberhart. Empirical Study of Particle Swarm Optimization. *IEEE Int. Congress on Evolutionary computation*, :101–106, 1999

Biographies

Wahiba Khemiri received the Engineering and Master degrees in 2004 and 2005 respectively from National Engineering School of Tunis. From 2010 she registered in Ph. D. in Monastir Engineering School. Her specific research interests are in the area Optimized Control of Motor Drives.

Anis Sakly received the Ph. D. degree in Electrical Engineering in 2005 from National Engineering School of Tunis. He is currently a Professor at the Electrical Department, National Engineering School of Monastir. His research interests are in analysis, synthesis and implementation of intelligent control systems. Particularly, his current research interests include soft computing-based approaches applied in optimal control, signal and image processing, and renewable energy systems optimization.

Mohamed Faouzi Mimouni received the Ph. D. and University habilitation degrees in Electrical Engineering Department of Monastir Engineering School, Tunisia, in 1997 and 2004 respectively and is currently a Professor. His specific research interests are in the area Power Electronics, Motor Drives, Solar and Wind Power Generation.

L. Rmili, S. Rahmani and K. Al-Haddad

PWM Modulation Technique of Three-Phase Indirect Matrix Converter

Abstract: This paper presents a Sparse Indirect Matrix Converter (SIMC) topology, which can be proposed as an alternative to the direct matrix converter. This double stage converter configuration consists of a combination of two conventional converters through a fictitious intermediate floor without capacitive storage element. The first floor is a controlled rectifier directly connected to a second floor, which consists of a voltage source inverter traditionally used in many different industrial applications. Pulse Width Modulation (PWM) control strategy of matrix converters is aimed to generate the gate signals that guarantee the control of the semi-conductors in good conditions. It provides balanced output voltages with controllable frequency, a maximum value of the ratio and a displacement factor near unity regardless the load.

Keywords: PWM Control, Sparse Indirect Matrix Converter, Power Conversion.

1 Introduction

The Indirect Matrix Converter (IMC) is a modern direct converter of AC/AC electrical power without dc-link capacitor. It consists of a matrix of bi-directional switches arranged such that any input phase can be connected to any output phase at any point in time. It ensures bidirectional power flow between the network and the receiver load with a control of the output voltage amplitude and frequency [11]–[1]. So, matrix converter provides an adjustable input power factor and high quality sine waveform through to a matrix structure of bidirectional power switches in current and voltage, in each output phase is really to each input phase. Matrix converter has a remarkable interest since it has appeared in 1976 especially during the last decade. View the advantage of IMC compared to conventional converters such as cyclo-converter, dimmer and conventional converter (cascade: rectifier-capacitor-inverter) including:

- A wide range of operating frequency to the output voltage.
- A variable ratio between the output and input voltage can be maximized if possible.
- The decouple controlling of output voltage amplitude and frequency.
- A reduced total harmonic distortion for the input and output currents.
- Input and output current and voltage sine wives with an adjustable phase shift, so the ability to operate at unity power factor for any load.

L. Rmili, S. Rahmani and K. Al-Haddad: L. Rmili, ISET of Tunis, Tunisia, email: rmili_lazhar@yahoo.fr, S. Rahmani, ISTMT, University of Tunis, Tunisia, email: rsalem02@yahoo.fr, K. Al-Haddad, École de Technologie Supérieure (ETS), Montreal, QC, Canada, email: kamal@ele.etsmtl.ca

De Gruyter Oldenbourg, ASSD – Advances in Systems, Signals and Devices, Volume 7, 2018, pp. 35–46.
https://doi.org/10.1515/9783110470529-003

– Operation in all four quadrants.
– The absence of a large capacitor for filtering and energy storage, bulky, heavy and susceptible to failure, which reduces the cost and design of the converter.
– Operation at high temperature.
– Gain reliability.

All these advantages facilitate the integration of this new converter topology in several areas of industrial applications such as aerospace industries that have a great interest in this converter, marine propulsion industries, the electric drive variable speed machines, embedded systems and renewable energy field based wind and fuel cells [2]–[8].

In this paper a sparse indirect matrix converter is developed and results are obtained for an RL output load. Simulation results are discussed, and the performance of the sparse indirect matrix converter topology is therefore evaluated.

2 Dual-stage matrix converter configurations

The two-stage indirect matrix converter structure developed by J.W. Kollar has a major advantage such as the ability to minimize the number of power transistors. Five topologies based on IGBT switch and two topologies based on RB-IBGT. The structures are presented in Fig. 1. Table 1 shows the summary of the above-mentioned matrix converter topologies considering various points including number of components, power losses, control strategy complexity and reversibility [1]–[4].

3 Sparse matrix converter

The configuration shown in Fig. 2, leads to remove an IGBT from each arm of the rectifier, so three components will be eliminated in all, compared to the previous configuration which facilitates the development of the control of the converter.

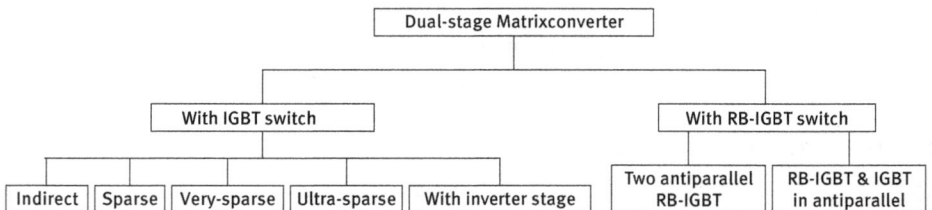

Fig. 1. Different structures of indirect matrix converter topologies.

Tab. 1. Summary of the properties of indirect topologies.

Topology	Number of transistor	Number of diodes	Energetic losses	Reversibility power	Control
Indirect matrix converter	18	18	Low	Yes	fairly complicated
Sparse	15	18	Important	Yes	fairly complicated
Very-Sparse	12	12	Low	Yes	easy
Ultra-Sparse	9	18	Low	No	easy
With stage inverter	14	14	Important	Yes	Complicated
Based on RB-IGBT	18	18	Low	Yes	fairly complicated
Based hybrid switches	18	18	Average	Yes	easy

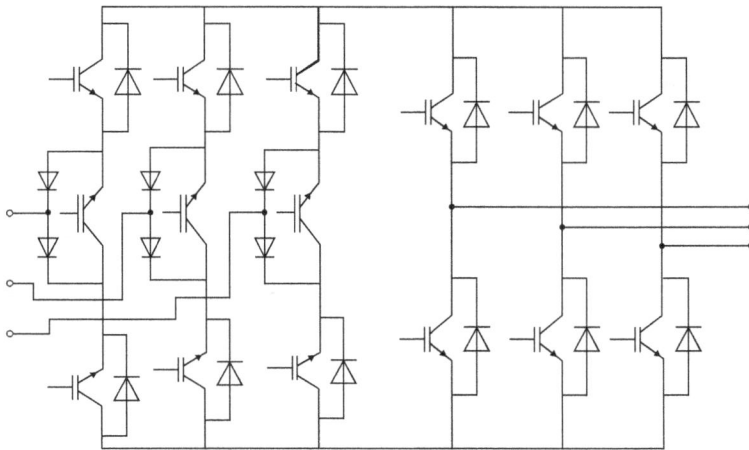

Fig. 2. Sparse indirect matrix converter topology.

Conduction losses will be greater than those generated by the first configuration since three transistors and diodes are three drivers during the feeding phase of the load also two transistors and two diodes in the feedback phase power to the network [1]–[7].

4 Modeling and principle operating of sparse indirect matrix converter

The principal indirect matrix converter topology is composed of two stages (rectifier and inverter) as shown in figure 3.

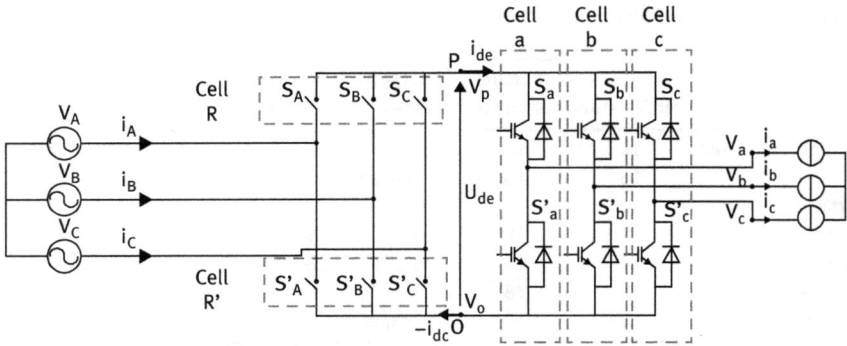

Fig. 3. Indirect matrix converter "dual-stage".

The rectifier stage is formed by two switching cells, denoted (R) and (R'). It is modeled by:

$$[S_{rect}] = \begin{bmatrix} S_A & S_B & S_C \\ S'_A & S'_B & S'_C \end{bmatrix} \tag{1}$$

where S_A, S_B, S_C are the state of the switches and $[S_{rect}]$ is the connection matrix of the rectifier. For each cell, one switch is closed at any switching time; this condition is expressed by:

$$\begin{cases} S_A + S_B + S_C = 1 \\ S'_A + S'_B + S'_C = 1 \end{cases} \tag{2}$$

The input currents and output voltages of the rectifier stage are expressed as:

$$\begin{bmatrix} v_p \\ v_o \end{bmatrix} = [S_{rect}]^T \cdot \begin{bmatrix} v_a \\ v_b \\ v_c \end{bmatrix} \tag{3}$$

$$\begin{bmatrix} i_A \\ i_B \\ i_C \end{bmatrix} = [S_{rect}]^T \cdot \begin{bmatrix} i_{dc} \\ -i_{dc} \end{bmatrix} \tag{4}$$

The inverter stage of the indirect matrix converter consists of three switching cells, called a, b, c as shown in Fig. 1. The inverter stage is modeled by equation (5) and satisfies the constraints described by (6).

$$[S_{inv}] = \begin{bmatrix} S_a & S'_a \\ S_b & S'_b \\ S_c & S'_c \end{bmatrix} \tag{5}$$

where $[S_{inv}]$ is the connection matrix of the inverter stage.

$$\begin{cases} S_a + S_a' = 1 \\ S_b + S_b' = 1 \\ S_c + S_c' = 1 \end{cases} \tag{6}$$

The inverter operation is determined by the relations (7) and (8):

$$\begin{bmatrix} V_a \\ V_b \\ V_c \end{bmatrix} = [S_{rect}] \cdot \begin{bmatrix} V_p \\ V_o \end{bmatrix} \tag{7}$$

$$\begin{bmatrix} i_{dc} \\ -i_{dc} \end{bmatrix} = [S_{rect}]^T \cdot \begin{bmatrix} i_a \\ i_b \\ i_c \end{bmatrix} \tag{8}$$

The connection matrix of two-stage matrix converter named $[S_{DE}]$ is obtained by the product of the connecting matrices of the inverter and rectifier, as shown in equation (9):

$$[S_{DE}] = [S_{inv}] \cdot [S_{rect}] = \begin{bmatrix} S_a & S_a' \\ S_b & S_b' \\ S_c & S_c' \end{bmatrix} \cdot \begin{bmatrix} S_A & S_B & S_C \\ S_A' & S_B' & S_C' \end{bmatrix} = \begin{bmatrix} S_{Aa} & S_{Ba} & S_{Ca} \\ S_{Ab} & S_{Bb} & S_{Cb} \\ S_{Ac} & S_{Bc} & S_{Cc} \end{bmatrix} \tag{9}$$

The conversion function of the sparse indirect matrix converter is established by equation (10):

$$[S] = [S_{DE}] \Rightarrow \begin{bmatrix} S_{Aa} & S_{Ba} & S_{Ca} \\ S_{Ab} & S_{Bb} & S_{Cb} \\ S_{Ac} & S_{Bc} & S_{Cc} \end{bmatrix} = \begin{bmatrix} S_a & S_a' \\ S_b & S_b' \\ S_c & S_c' \end{bmatrix} \cdot \begin{bmatrix} S_A & S_B & S_C \\ S_A' & S_B' & S_C' \end{bmatrix} \tag{10}$$

In the same manner as the direct matrix converter, a formulation based on modulation of the switches may also be set for the sparse indirect matrix converter. The equations described above in "connection function" are transformed in "modulation function" and the conversion matrices defined by the modulation functions of each stage of indirect matrix converter are described by equations (12) for the rectifier stage and by (16) for the inverter stage.

If t_i represents the conduction time of the corresponding switch S_i during the switching period T, then the modulation coefficients relative to the conduction time is expressed by equation (11):

$$m_i = \frac{t_i}{T} \tag{11}$$

Where $i = \begin{cases} A \\ B \\ C \end{cases}$ for the rectifier and $\begin{cases} a \\ b \\ c \end{cases}$ for the inverter.

$$[M_{rect}] = \begin{bmatrix} m_A & m_B & m_C \\ m'_A & m'_B & m'_C \end{bmatrix} \tag{12}$$

The modulation coefficients must satisfy the following conditions (13):

$$\begin{cases} m_A + m_B + m_C = 1 \\ m'_A + m'_B + m'_C = 1 \end{cases} \tag{13}$$

The relationship between the instantaneous values of the output and input voltages is then given by equation (14):

$$\begin{bmatrix} v_p \\ v_o \end{bmatrix} = [M_{rect}] \begin{bmatrix} V_A \\ V_B \\ V_C \end{bmatrix} \tag{14}$$

The input and output current of the rectifier stage are given by equation (15):

$$\begin{bmatrix} i_A \\ i_B \\ i_C \end{bmatrix} = [M_{rect}]^T \begin{bmatrix} i_{red} \\ -i_{red} \end{bmatrix} \tag{15}$$

$$[M_{inv}] = \begin{bmatrix} m_a & m'_a \\ m_b & m'_b \\ m_c & m'_c \end{bmatrix} \tag{16}$$

The modulation coefficients must satisfy the following conditions (17):

$$\begin{cases} m_a + m'_a = 1 \\ m_b + m'_b = 1 \\ m_c + m'_c = 1 \end{cases} \tag{17}$$

The relationship between the instantaneous values of the output and input voltages is then given by equation (18):

$$\begin{bmatrix} v_a \\ v_b \\ v_c \end{bmatrix} = [M_{inv}] \begin{bmatrix} v_p \\ v_o \end{bmatrix} \tag{18}$$

Similarly the input and output current of the inverter stage are expressed by equation (19):

$$\left[\begin{array}{c} i_{dc} \\ -i_{dc} \end{array} \right] = [M_{inv}]^T \left[\begin{array}{c} i_a \\ i_b \\ i_c \end{array} \right] \tag{19}$$

The product conversion matrix of the inverter stage and the rectifier stage gives the conversion matrix $[M_{DE}]$ which is expressed by equation (20):

$$[M_{DE}] = [M_{inv}].[M_{rect}] = \left[\begin{array}{cc} m_a & m'_a \\ m_b & m'_b \\ m_c & m'_c \end{array} \right] . \left[\begin{array}{ccc} m_A & m_B & m_C \\ m'_A & m'_B & m'_C \end{array} \right] \tag{20}$$

5 PWM control of sparse indirect matrix converter topology

The PWM control technique is used to calculate the duty cycle of the switches which is inspired by comparison of a reference sine wave with a triangular wave in order to generate pulses with modulation varied duty cycle according to the frequency of the reference sine wave. The PWM control the energy supply to the load feeding by matrix converter [3]–[13].

6 Simulations results

The simulation results of the sparse matrix converter feeding an $R - L$ load (R = 10 Ω, $L = 10$ mH), before and after compensation are shown in Figs. 4 to 9. The SIMC system has been simulated using the "Power system Blockset" simulator operating under Matlab/Simulink environment. Figure 4 shows the input current i_e(A) and it's harmonic spectrum. The total harmonic distortion (THD) of input current is 77 %. The output voltage v_s (V), it's harmonic spectrum and the sinusoidal output waveform current i_s (A) are shown in Figure 5 and 6 respectively. The THD of the output voltage is 92.22 %.

An input $L - C$ filter (L_i = 40 mH, C_i = 100 nF) and output $L - C$ filter (L_o = 27 mH, Co = 220 μF) are designed and used to compensate harmonics and reactive power of the output/input voltages/currents. Figure 7 shows the sinusoidal waveforms with high quality of input current. The THD of the input current has been reduced from 77 % before filtering to 1.87 % after filtering. Figure 8 shows the high quality of sinusoidal

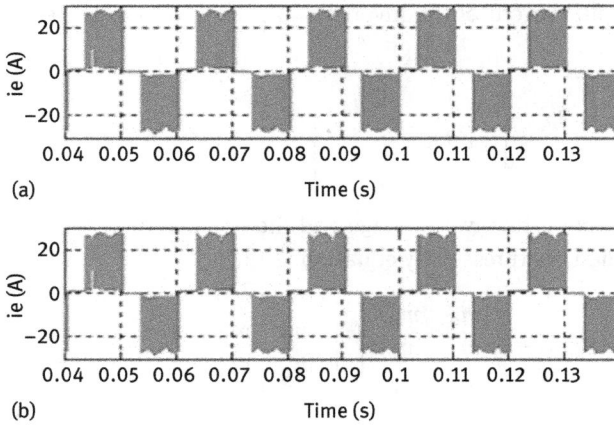

Fig. 4. (a) Simulation results of the input current ie of three-phase sparse matrix converter before filtering (b) Spectrum of the input current.

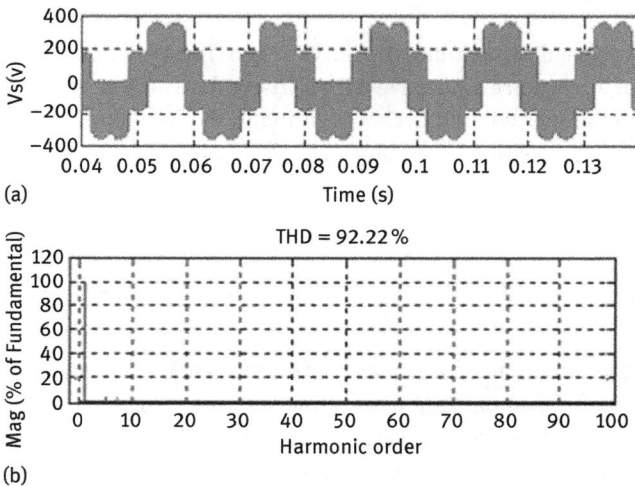

Fig. 5. (a) Simulation results of the output voltage v_s of three-phase sparse matrix converter before filtering, (b) Spectrum of the output voltage.

waveforms of the output current (THD = 0.26 %). The output voltage is (THD = 0.36 %) shown in Fig. 9. These results show that the ratio q of the output voltage from the input voltage reaches its maximum value 0.852.

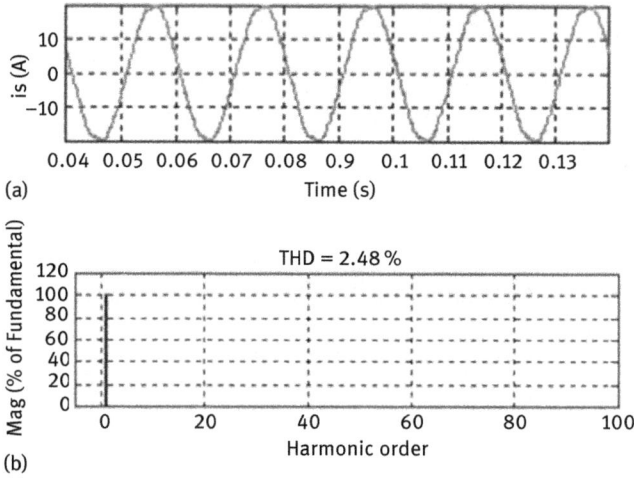

(a)

(b)

Fig. 6. (a) Simulation results of the output current i_e of three-phase sparse matrix converter before filtering, (b) Harmonic spectrum of the output current.

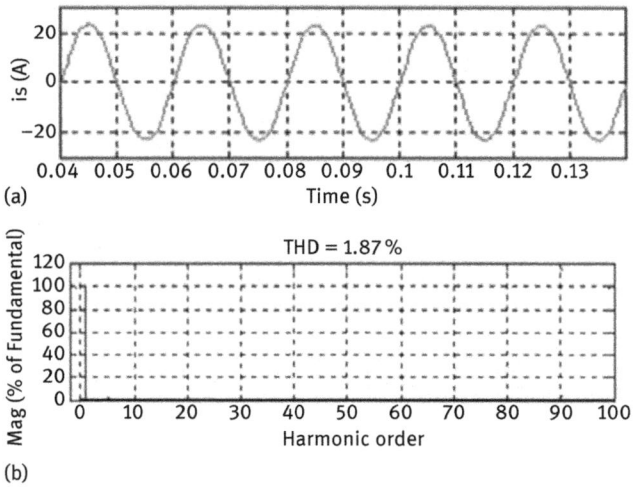

(a)

(b)

Fig. 7. (a) Simulation results of the input current i_e of three-phase sparse matrix Converter after filtering, (b) Spectrum of the input current.

7 Conclusion

In this paper a sparse topology of indirect matrix converter is developed and results are obtained for an RL output load. Further, the operation of a three phase to three phase

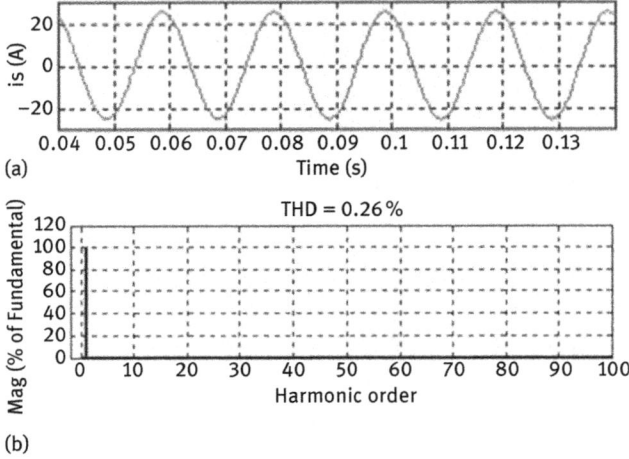

Fig. 8. (a) Simulation results of the output current of three-phase sparse matrix converter after filtering, (b) Spectrum of the output current.

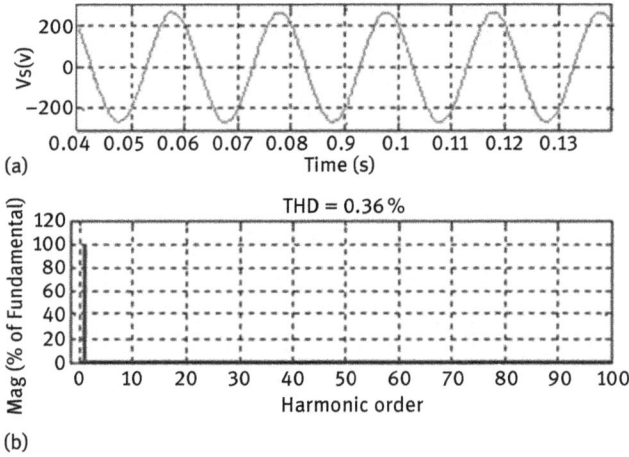

Fig. 9. (a) Simulation results of the output voltage v_s of three phase sparse matrix converter before filtering, (b) Spectrum of the output voltage.

matrix converter synthesized using the PWM control has been detailed. This paper is helpful to study different topology of matrix converter with many other different loading and synthesize the input current and output voltage using various control technique. It offers a very wide field of research especially in the study of reliability, maintainability, availability; faults tolerances; stability of these types of converters and the possibility of drawing developed strategies controls ensuring the optimization of conduction and switching losses with high performance operation of the converter. The matrix converter has advantages:

- The possibility of reducing the number of switches forming the converter thus reducing the number of semi-conductors and subsequently losses.
- These difficulties switching are reduced, switches of the input stage (rectifier) are reduced and the second stage of the converter switches as a standard inverter.

Bibliography

[1] R.L. A Ribeiro T.M. Oliveira E.R.C. Da Silva C.B. Jacobina R. Correa and A.M.N. Lima. AC/AC converters with a reduced number of switches. *IEEE IAS Industry Applications Society Annual Meeting*, :1755–1762, 2001.

[2] I. Boldea C. Klumpner, P. Nielsen and F. Blaabjerg. A new matrix converter motor (mcm) for industry applications. *IEEE Trans. on Industrial Electronics*, 49:325–335, April, 2002.

[3] J. Choi and S. Sul. A new compensation strategy reducing voltage/current distortion in pwm vsi systems operating with low voltages. *IEEE IAS Annual Meeting*, 31:1001–1008, 1995.

[4] F.A.S. Neves F. Bradaschia, M. Cavalcanti and E.P. Helber. A modulation technique to reduce switching losses in matrix converters. *IEEE Trans. on Industrial Electronics*, 56(4):1186–1195, April, 2009.

[5] A. Garces and M. Molinas. A study of efficiency in a reduced matrix converter for offshore wind farms. *IEEE Trans. on Industrial Electronics*, 59(1):184–193, January, 2012.

[6] M. Ikeda J. Kang, E. Yamamoto and E. Watanabe. Medium voltage matrix converter design using cascaded single phase power cell modules. *IEEE Trans. on Industrial Electronics*, 58(11):5007–5013, November, 2011.

[7] J. Weigold J. Mahlein and O. Simon. New concepts for matrix converter design. 27th *Annual Conf. of the IEEE Industrial Electronics Society IECON*, 2:1044–1048 November 29, December 2, 2001.

[8] C. Klumpner and F. Blaabjerg. Using reverse blocking igbts in power converters for adjustable speed drives. *IEEE Industry Applications Conf.*, 3:1516–1523, October, 2003.

[9] F. Fnaiech L. Rmili, S. Rahmani and K. Al Haddad. Space vector modulation strategy for a direct matrix converter. 14th *Int. Conf. on Sciences and Techniques of Automatic Control and Computer Engineering* (STA), :126–131, 2013.

[10] H. Vahedi L. Rmili, S. Rahmani and K. Al Haddad. A comprehensive analysis of matrix converters: Bidirectional switch, direct topology, modeling and control. 23rd *IEEE Int. Symp. on Industrial Electronics* (ISIE), :313–318, 2014.

[11] J. Clare L. Empringham P. Wheeler, J. Rodriguez and A. Weinstein. Matrix converters : A technology review. *IEEE Trans. Industrial Electronics*, 49(2):276–288, April, 2002.

[12] J. Rodriguez. High performance dc motor drive using a pwm rectifier with power transistors. *Inst. Elect. Eng. B Elect. Power Appl.*, 134(1):9–13, January, 1987.

[13] T. Yamashita and T. Takeshita. PWM strategy of single-phase to threephase matrix converters for reducing a number of commutations. *IPEC*, :3057–3064, June, 2010.

Biographies

Lazhar Rmili was born in Kasserine, Tunisia. He received the B. Sc. A. and M. Sc. A. (electrical) degrees and the Specialized Scientific Studies Certificate (CESS) from the High School of Sciences and Technologies of Tunis (ESSTT), Tunis, Tunisia, in 2000, 2008, and 2004, respectively, and the highest academic-rank qualification (Agrégation) degree in Electrical Engineering in 2006. In September 2006, he was an Enseignant Technologue with the Department of Electrical Engineering, High Institute of Technology's study of Rades, Tunis (ISET) Rades, Tunisia. His fields of interest include power quality, active filters, matrix converter, and multilevel converter, including power converter topology, modeling, and control aspects.

Salem Rahmani was born in Tunisia. He received the B. Sc. A. and M. Sc. A. (electrical) degrees and the Specialized Scientific Studies Certificate (CESS) from the High School of Sciences and Technologies of Tunis (ESSTT), Tunis, Tunisia, in 1992, 1995, and 2001, respectively, and the highest academic-rank qualification (Agrégation) degree in Electrical Engineering in 2001, and the Ph. D. and the highest academic qualification (Habilitation qualifications (HDR)) degrees from the National Engineering School of Tunis (ENIT), Tunis, Tunisia, in 2004 and 2010, respectively. In September 2002, he was an Assistant Professor with the Department of Electrical Engineering, High Institute of Medical Technologies of Tunis (ISTMT), Tunis, Tunisia. Since the elaboration of his Ph. D. degree, he has been a Member of the Research Group in Power Electronics and Industrial Control (GREPCI), école de Technologie Supérieure, University of Québec, Montreal, QC, Canada. His fields of interest include power quality, active filters, multilevel converter, and resonant converters, including power converter topology, modeling, and control aspects.

Kamal Al-Haddad received his B. Sc. A. and M. Sc. A. degrees from the university of Québec Ã Trois-Rivières, Trois-Rivières, QC, Canada, in 1982 and 1984, respectively, and his Ph. D. from the Institut National Polytechnique, Toulouse, France, in 1988. From June 1987 to June 1990 he was a Professor with the Engineering Department, Université du Québec Ã Trois Rivières. Since June 1990, he has been a Professor with the Electrical Engineering Department, école de Technologie Supérieure (ETS), Montreal, QC, where he has been the holder of the Canada Research Chair in Electric Energy Conversion and Power Electronics since 2002. He has supervised more than 100 M. Sc. A. and Ph. D. students working in the field of power electronics. He was the Director of graduate study programs at ETS from 1992 to 2003. His fields of interests include high-efficient static power converters, harmonics and reactive power control using hybrid filters, and switch-mode and resonant converters, including the modeling, control, and development of prototypes for various industrial applications in electric traction, power supply for drives, telecommunication, etc. He is very active in the industrial Electronics Society where he is president elect, AdCom Member and served as an Associate editor for IEEE Transactions on Industrial Electronics and Industrial Informatics.

S. Mabrouk, A. Bouallegue and A. Khedher

Modeling of Unbalanced Radial Distribution System and Backward – Forward Power Flow Analysis

Abstract: Electrical energy systems are undergoing a radical transformation in structure and functionality in a quest to increase efficiency and reliability. Due to these changes power systems became larger and larger, transmission lines crisscrossed the land forming large interconnected network. It becomes very important and necessary to operate a distribution system at its maximum capacity. For these reasons distribution system has to be modeled accurately so that analysis techniques for steady state and short circuit conditions can be developed [1]. In this work, a contribution for a modeling of a three phased unbalanced radial distribution system components, is given. A decoupled approach has been adopted so that each phase can be studied independently of the others. Some important components of a simple radial feeder has been described such as the line segment, the voltage regulator, and different models of loads have been presented and discussed. The obtained model has been tested for a backward/forward power flow analysis applied on an IEEE 69 bus radial distribution system.

Keywords: Modeling, Radial, Distribution System, Three-Phase, Unbalanced

1 Introduction

Distribution systems serve as the link from the distribution substation to the customer. It is the core of modern smart grid and the provider of safe and reliable energy to different areas. The Distribution feeder circuits usually consist of overhead and underground circuits in a mix of branching laterals from the station to the various customers. The circuit is designed around various requirements such as required peak load, voltage, distance to customers, and other local conditions such as terrain, visual regulations, or customer requirements [2]. The various branching laterals of a distribution system are usually in a radial configuration so that energy's propagation is in one direction from the source to the load. For smart grid the integration of distributed energy has created more complication in the power system and the analysis of the

S. Mabrouk, A. Bouallegue and A. Khedher: S. Mabrouk, High Institute of technology of Sousse, Tunisia, email: mabrouk_souhir@yahoo.com, A. Bouallegue, National Engineering school of Sousse, Tunisia, email: adel.bouallegue@eniso.rnu.tn, A.Khedher, National Engineering school of Sousse, Tunisia, email: adel_kheder@yahoo.fr

De Gruyter Oldenbourg, ASSD – Advances in Systems, Signals and Devices, Volume 7, 2018, pp. 47–60.
https://doi.org/10.1515/9783110470529-004

distribution system under these conditions has become a difficult. In fact, electric power distribution systems are characterized by low R/X ratios and unbalanced operations. These characteristics impose serious challenges for the development of efficient computational power flow techniques [3]. For this purpose a good and accurate model is needed. Diverse papers have tried to model the components of the distribution feeder: Peng Xiao [4], T.H Chen [5], Baran in [6], and Wang in [7] have tried to propose a unified model of the transformer in at distribution system based on the computing of primary admittance matrix and a nodal admittance matrix of the transformer. W. H. Kersting in [8] has tried to propose a complete model for all the components of a distribution system. In this work, some of the components of three phased radial distributed feeder has been modeled, such as lines, voltage regulator and loads; the proposed models have been used in a modified IEEE 69 nodes radial system to compute a power flow analysis based on a backward/forward algorithm, the obtained convergence velocity of the algorithm proves the efficiency of the models.

2 Distribution Components Modeling

2.1 A simple distribution feeder components

A substation serves one or more feeders; there are some usual components that are composing any distribution system such as transformers, voltage regulators, lines, capacitor shunt banks, etc. An example of a simple representation of a feeder is shown in Fig. 1. In a radial feeder, the power flow has one direction from the source substation to each customer. Distribution feeders are inherently unbalanced due to the large numbers of unequal loads connected to the grid and the different spacing between three-phase overhead and underground conductors. A distribution feeder must be correctly mapped so that information on lines, distances, locations, connections and KVA rating can be explicit.

2.2 Three-phase line modeling

High voltage transmission lines are transposed and supposed to be balanced while distribution systems are serving unbalanced lines and have untransposed, mutual and self-impedances that have to be developed in the case of three- phase system for each phase. The following modified Carson's equations (1) and (2) are used to compute mutual and self line impedance. Results are usually given in /mile but in this paper based on equations given in [8], and [3] results will be in km for phases i and j.

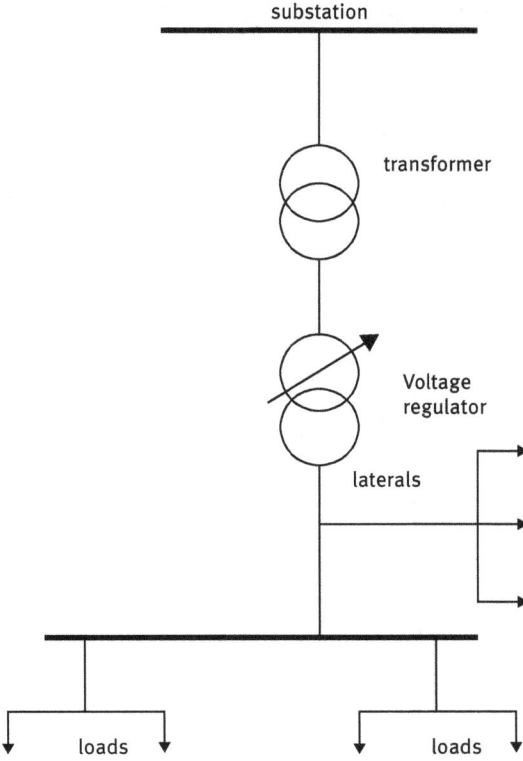

Fig. 1. Simple distribution feeder components.

$$z_{ii} = r_i + 0.05919 \times \left(\frac{f}{50}\right) + j \times 0.07537 \left(\frac{f}{50}\right) \times \left[\ln\left(\frac{1}{GMR_i}\right) + 6.74580\right] \quad (1)$$

$$z_{ij} = r_i + 0.05919 \times \left(\frac{f}{50}\right) + j \times 0.07537 \left(\frac{f}{50}\right) \times \left[\ln\left(\frac{1}{D_{ij}}\right) + 6.74580\right] \quad (2)$$

$$D_{ij} = GMD = (D_{ab} + D_{bc} + D_{ca})^{\frac{1}{3}} \quad (3)$$

where: r_i is the resistance of conductor in Ωkm, f is the nominal frequency of the grid (50 hz), GMR_i is the geometric mean radius of the conductor i, Z_i is the line to neutral load impedance, Z_{ij} is the line to line load impedance, D_{ij} is the distance between phases i and j and it is called also geometric mean distance between phases GMD GMD is computed for phases a, b and c. Appliance of (1) and (2) to the three-phases segment of line gives the primitive impedance matrix as follows:

$$z_{prim} = \begin{bmatrix} z_{aa} & z_{ab} & z_{ac} & z_{an} \\ z_{ba} & z_{bb} & z_{bc} & z_{bn} \\ z_{ca} & z_{cb} & z_{cc} & z_{cn} \\ z_{na} & z_{nb} & z_{nc} & z_{nn} \end{bmatrix} \quad (4)$$

This matrix can be reduced to a 3×3 matrix using Kron reduction [1], and [11], the following equation is then used:

$$z'_{ij} = z_{ij} - z_{in} \times \left(\frac{z_{nj}}{z_{nn}}\right) \tag{5}$$

The obtained impedance matrix is then:

$$z'_{abc} = \begin{bmatrix} z'_{aa} & z'_{ab} & z'_{ac} \\ z'_{ba} & z'_{bb} & z'_{bc} \\ z'_{ca} & z'_{cb} & z'_{cc} \end{bmatrix} \tag{6}$$

2.3 Voltage regulator

Usually the most used is three-phases step voltage regulator which can be connected only in Wye or closed delta mode. A step voltage regulator consists of an auto-transformer and a load tap changing mechanism, the voltage change is obtained by changing the taps of the series windings of the autotransformer. Step regulator can be connected in a type A or type B according to the ANSI IEEEC57.15–1986 standard. We'll be first modeling the type B for it is the most commonly used. Figure 2 gives the Type B step regulator in the raise position.

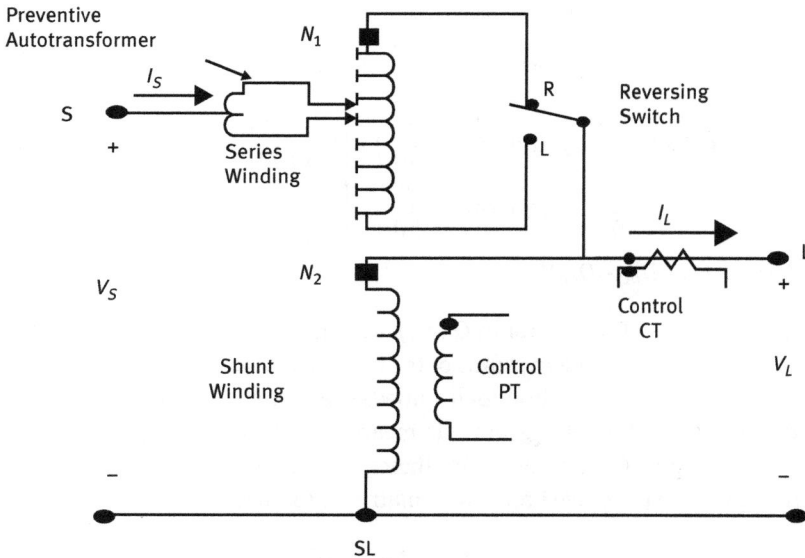

Fig. 2. Type B step voltage regulator in the raise position [9].

The voltage equations and current equations in the raise and lower positions (N1 and N2 are the numbers of primary and secondary windings) are given in [1]. For the three type B single-phase regulator Wye connected, considering A, B, and C the source side of the regulator, and a, b, and c the load side phasing, W.H. Kirsting in [1], and in [9] has presented the three phased model as follow: The Voltage equations are:

$$
\begin{bmatrix} V_{An} \\ V_{Bn} \\ V_{Cn} \end{bmatrix} = \begin{bmatrix} a_{Ra} & 0 & 0 \\ 0 & a_{Rb} & 0 \\ 0 & 0 & a_{Rc} \end{bmatrix} \times \begin{bmatrix} V_{an} \\ V_{bn} \\ V_{cn} \end{bmatrix}
\tag{7}
$$

where V_{An}, V_{Bn} and V_{Cn} are the voltage regulator load side line to neutral voltages and a_{Ra}, a_{Rb} and a_{Rc} are the voltage regulator source side line to neutral voltages. The Current equations are:

$$
\begin{bmatrix} I_A \\ I_B \\ I_C \end{bmatrix} = \begin{bmatrix} \frac{1}{a_{Ra}} & 0 & 0 \\ 0 & \frac{1}{a_{Rb}} & 0 \\ 0 & 0 & \frac{1}{a_{Rc}} \end{bmatrix} \times \begin{bmatrix} I_a \\ I_b \\ I_c \end{bmatrix}
\tag{8}
$$

where: I_a, I_b and I_c are the voltage regulator load side currents and I_A, I_B and I_C are the voltage regulator source side currents. The voltage regulator is generally put after the swing bus or the main transformer at the source side of the network. For each of phases a, b or c, a_R is the rate of the voltage regulator, it is computed for a type B step voltage regulator in a raise position as follow:

$$
a_R = 1 - \frac{N_1}{N_2}
\tag{9}
$$

For type B step voltage regulator in a lower position

$$
a_R = 1 + \frac{N_1}{N_2}
\tag{10}
$$

Using these values of voltages and currents at the load side can be easily deduced from (7) and (8) by inverting the given matrices. Figure 3 gives a type A step-voltage regulator in the raise position, the primary circuit of the system is connected directly to the shunt winding of the Type A regulator. The series winding is connected to the shunt winding and, in turn, via taps, to the regulated circuit. In this connection the core excitation varies because the shunt winding is connected directly across the primary circuit.

For both types A and type B regulators, the relationship between the source voltage and the current (indexed by s) to the load voltage and current (indexed by L) are:

$$
V_s = \frac{1}{a_R} \times V_L, \qquad I_s = a_R \times I_L
\tag{11}
$$

$$
V_s = a_R \times V_L, \qquad I_s = \frac{1}{a_R} \times I_L \quad !!!!
\tag{12}
$$

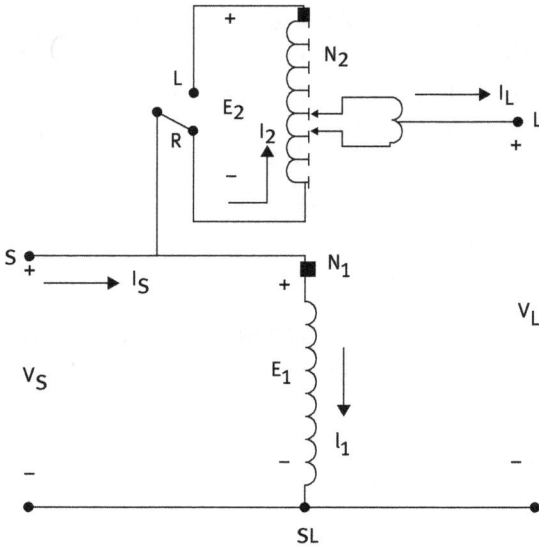

Fig. 3. A step voltage regulator in the raise position [9].

2.4 Load models

Loads vary continually with a high degree of uncertainties caused by the dependency of loads on human activities and temporal factors. There are several models of three phase unbalanced loads:

- Constant real and reactive power (constant PQ)
- Constant current
- Constant impedance
- Any combination of the above

In a distribution feeder loads can be Wye-connected or Delta-connected.

2.4.1 Wye-connected loads

For the three phases a, b and c we will use the next notation: for phase i where , $i \in a$, b and c, θ_i is the power factor angle of line i, δ_i is the line i to neutral voltage angle, S_i the complex line to neutral load power and V_{in} voltage's line to neutral are:

$$|S_i|\angle\theta_i = P_i + j \times Q_i \tag{13}$$

$$V_{in}\angle\delta_i \tag{14}$$

2.4.1.1 Real and reactive power (constant PQ)

The line current for a phase i, where

$$I_i = \left(\frac{S_i}{V_{in}} \right)^* \tag{15}$$

In the iterative process of power flow calculation will change until achieving convergence.

2.4.1.2 Constant impedance loads

The constant load impedance Z_i for phase i, where $i \in a, b$ and c is:

$$Z_i = \frac{|V_{in}|^2}{S_i^*} \tag{16}$$

Then the load current can be computed as follow:

$$I_i = \frac{V_{in}}{Z_i} \tag{17}$$

In the iterative process of power flow calculation $V_{i}n$ will change until achieving convergence, but Z_i will remain constant.

2.4.1.3 Constant current loads

In the iterative process of power flow calculation:
- first step: computing the magnitude of currents by the equation (15).
- second step: the current are held constant while the line to neutral voltage angle changes.

This makes the current angle change so that the power factor of the load remains constant.

2.4.2 Delta-connected Loads

The notations used in this section are as follow: for phase ij where $ij \in ab, bc$ or ac, θ_{ij} is the power factor angle between lines i and j, δ_{ij} is the line i to line j voltage angle, S_{ij} is complex line to line load power:

$$|S_{ij}| \angle \theta_{ij} = P_{ij} + j \times Q_{ij} \tag{18}$$

$$V_{ij} \angle \delta_{ij} \tag{19}$$

2.4.2.1 Real and reactive power (constant PQ)

The line current between phases i and j, where :

$$I_{ij} = \left(\frac{S_{ij}}{V_{ij}} \right)^{*} \tag{20}$$

In the iterative process of power flow calculation the line to line voltage will change until achieving convergence.

2.4.2.2 Constant impedance loads

The constant load impedance Z_{ij} for phase i and j, where $ij \in ab$, bc or ca, is:

$$Z_{ij} = \frac{|V_{ij}|^2}{S_{ij}^{*}} \tag{21}$$

Then the load current between lines I_{ij} can be computed as follow:

$$I_{ij} = \frac{V_{ij}}{Z_{ij}} \tag{22}$$

In the iterative process of power flow calculation, V_{ij} will change until achieving convergence but Z_{ij} will remain constant.

2.4.2.3 Constant current loads

In the iterative process of power flow calculation:
- First step: computing the magnitude of currents by the equation (20).
- Second step: the current are held constant while δ_{ij} the line to line voltage angle changes.

This makes the current angle changes so that the power factor of the load remains constant.

3 Simulation example

The modeling of the distribution system is the first step to compute network analysis, one important way to achieve that is to compute steady state power flow analysis. The IEEE test cases are good examples to test the efficiency of such an algorithm. A modified IEEE 69 nodes radial distribution feeder has been used to compute a backward/forward power flow equations, using the described models. Only the topography of the grid has been conserved, line characteristics, load parameters, the position of voltage regulators, and transformers are proposed. The described algorithm has been used by Liu and al. [15] at first with balanced radial or meshed networks. Backward/forward unbalanced distribution system power flow analysis was given by Teng [16] for a three phases grid. The chart in Fig. 4. gives the steps of the proposed algorithm.

– P_{ij}: real power between buses i and j
– Q_{ij}: reactive power between buses i and j
– $ACSR$: Aluminum Conductor Steel reinforced
– V_{abcl}: voltage line matrix for phases a, b and c

Figure 5. gives the IEEE 69 node test case. The swing bus is chosen to be the bus 1. The lines are supposed to be ACSR conductors (Bare Aluminum Conductor Steel Reinforced), we chose the same conductor for phases and neutral. The conductor lines are composed of three phases and a neutral.

Table 1 gives the main characteristics of the conductor, [14]. The geometric mean distance GMD has been calculated for each phase a, b and c using (3), for the following

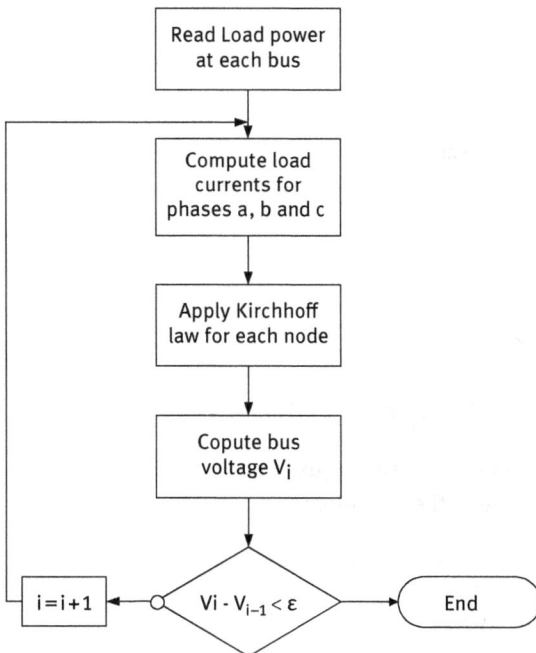

Fig. 4. Steps of the performed algorithm.

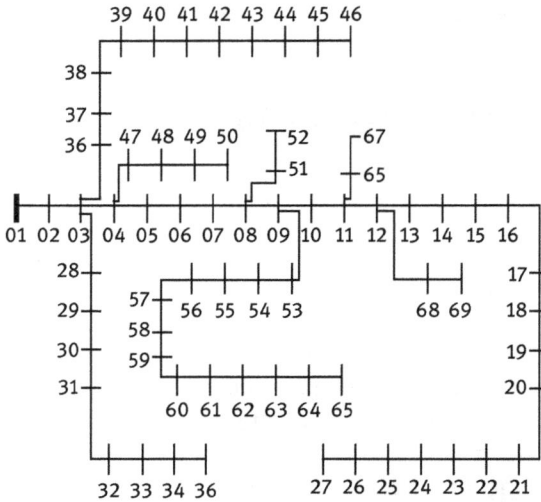

Fig. 5. IEEE 69 bus test case.

Tab. 1. Conductor characteristics.

Diameter of the conductor	97 mm
DC resistance at 20° C	0.0894 Ω/km
Calculated breaking load	135.13 kN
Nominal area of complete conductor	400 mm²

values:

$$D_{ab} = 0.762\,m \quad D_{bc} = 1.3716\,m \quad D_{ca} = 2.1336\,m$$
$$D_{an} = 1.7242\,m \quad D_{bn} = 1.3021\,m \quad D_{ca} = 1.524\,m$$

The impedances of the branches are computed using equations (1) and (2), the swing bus current is given, the obtained voltage at the first line is:

$$V_{abcl}(kV) = \begin{bmatrix} 1.7130 + 4.5401 \times i \\ 1.7889 + 1.3478 \times i \\ -0.6914 + 2.130 \times i \end{bmatrix}$$

where : V_{abcl} is voltage line matrix for phases a, b and c.

Tab. 2. PQ load Current.

Source node	Destination node	PQ load current (KA)		
		Ia	Ib	Ic
1	2	0.521−0.0104i	0.781−0.560i	0.5145+0.343i
2	3	0.647−0.001i	0.3794−0.274i	0.7832+0.542i
3	4	0.728−0.0027i	0.825−0.606i	0.342+0.236i
4	5	0.8579−0.0129i	0.322−0.2431i	0.637+0.446i
5	6	0.5383−0.0142i	0.228−0.172i	0.688+0.477i
6	7	0.670−0.007i	0.1938−0.161i	0.4238+0.295i
7	8	0.487−0.017i	0.439−0.335i	0.3461+0.2328i

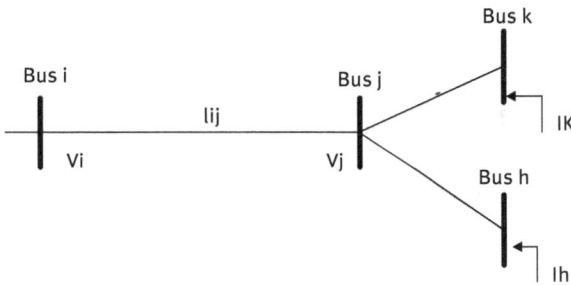

Fig. 6. Example of node of the radial feeder.

A three phase type B step voltage regulator in the raise position is put at the line 1 from bus 1 to bus 2 Wye grounded. The obtained voltage at bus 2 is then:

$$V_{abcreg}(V) = \begin{bmatrix} -84.17 - 223.09 \times i \\ -791.13 - 596.03 \times i \\ 305.78 - 942.15 \times i \end{bmatrix}$$

For each line in the system we have assigned a load, the computing of the load current is then done for the three types of the loads:
- PQ constant
- Impedance constant
- Current constant

The connection of the loads used in this example is the Wye connection. Some of the obtained results are shown in table 3.

Tab. 3. Computed voltages for phase A by the forward algorithm.

Bus n°	$V_{ja}(V)$	$V_{jb}(V)$	$V_{jc}(V)$
1	−1406.66+0.00i	−10307.53−7350.53i	−10307.5+7350.530i
2	4162.92+7728.036i	8512.39−2142.80i	7491.22+4575,365i
4	−8541.64+2023.08i	−8756.55−612.49i	8692.631+1220.91i
5	−8525.429−2090,34i	8690.47−1236.16i	−7356.55+4788.89i
6	556.52−8760.29i	8370.22−2644.21i	8204.64−3120.29i
7	−8777.59−79,5701i	−8623.18−1641.08i	212.55+8775.37i
8	2662.84+8364.31i	−5362.0+6949.911i	5049.29+7180.325i
9	8280.59−2912.77i	8250.01+2998.287i	−8612.53+1696.107i
10	−2529.76+8405.52i	3293.55+8136.64i	−2103.60+8522.16i

For the calculus of injection current at bus a backward approach [11], [13] has been adopted so that for each line ij:

$$I_{jk} = -I_k \tag{23}$$

$$I_{jh} = -I_h \tag{24}$$

$$I_{ij} = I_{jk} + I_{jh} - I_j \tag{25}$$

The next step is to compute the forward algorithm for the bus voltages for each branch between buses i and j:

$$V_j = V_i - Z_{ij} \times I_{ij} \tag{26}$$

The iteration n converges if the difference of bus voltages for consecutive iterations is equal or less to a given tolerance. For a bus j for the iteration n:

$$\Delta V_j^n = V_j^n - V_j^{n-1} \tag{27}$$

For phases a,b,c, for each bus j, ε is a given tolerance

$$\mathbb{R}(V_j^n) < \varepsilon \tag{28}$$

$$\mathbb{J}(V_j^n) < \varepsilon \tag{29}$$

If these equations are satisfied then the iteration stops, otherwise the process is repeated. If the load flow converges all the branch currents and voltages are known at each bus and the real and reactive power can therefore be calculated.

An example of the obtained values of are shown in Tab. 3, they are used to update the currents values until reaching the convergence criteria. In this paper the convergence of the applied algorithm has been obtained for a number of iterations equal to 8 for $\varepsilon = 3\%$. To evaluate the obtained result we execute the Deterministic Load Flow (DLF) Algorithm for the same IEEE 69 nodes example we found for $\varepsilon = 3\%$

a number of iteration equal to 6 which make it faster than the forward/backward algorithm, but it does not take in consideration the internal specificity of the grid such as cable nature and components like voltage regulators.

4 Conclusion

In this work, has been presented models for a simplified three phases radial unbalanced feeder, the model for each component is simple and decoupled for it can be used directly without the need of symmetrical components, besides, each phase can be used individually and the application of power flow algorithm is simplified. Three phase power flow for unbalanced radial distribution feeder has been computed using a backward/forward algorithm on a modified IEEE 69 node feeder. The calculus is simplified thanks to the models and the convergence criteria are achieved with acceptable velocity.

Bibliography

[1] W.H. Kersting. Distribution system modeling and analysis. *CRC press-Electric power engineering series*, USA 2002.

[2] J.D. McDonald, B. Wojszczyk, B. Flynn and I. Voloh. Distribution Systems, Substations, and Integration of Distributed Generation. *Electrical Transmission Systems and Smart Grids*, springer 2012.

[3] L. Uyttersprot. Short-circuit protection for combined overhead line-cable circuits. *IEEE-cable workshop*, Bern 2010.

[4] P. Xiao. A unified three-phase transformer model for distribution load flow calculation. *IEEE Trans. on power systems*, 21(1) Feb 2006.

[5] T.H. Chen, M.S. Chen, T. Inoue, P. Kotas and E.A. Chebli. Three phase cogenerator and transformer models for distribution system analysis. *IEEE Trans. Power Del.*, 6(4):1671–1681, Oct. 1991.

[6] M.E. Baran and E.A. Staton. Distribution transformer models for branch current based feeder analysis. *IEEE Trans. Power Syst.*, 12(2):698–703, May 1997.

[7] Z. Wang. Implementing Transformer Nodal Admittance Matrices Into BackwardForward Sweep-Based Power Flow Analysis for Unbalanced Radial Distribution Systems. *IEEE Trans on power systems*, 19(4):698–703, Nov. 2004.

[8] F. Calero. Mutual Impedance in Parallel Lines - Protective Relaying and Fault Location Considerations. *The Georgia Tech Protective Relaying Conference*, 2007.

[9] W.H. Kersting. Distribution Feeder Voltage Regulation Control. *Distribution Feeder Voltage Regulation Control*: pp. C1–C17, Fort Collins, 2009.

[10] W.H. Kersting. A Comprehensive Distribution TestFeeder. *IEEE Transmission and Distribution Conf. and Exposition PES*, New Orleans, April 2010.

[11] A. Ulinuha. Application of forward/backward propagation algorithm for three- power flow analysis in radial distribution system. *symposium Nasional RAPI VIII*, 2009.

[12] S. Bapu, C. Prakash and S.M. Kannan. Optimal Capacitor Allocation in 69-bus Radial Distribution System to Improve Annual Cost Savings for Dynamic Load. *Int. Journal of Emerging Technology and Advanced Engineering*, 3(3), March 2013.

[13] U. Eminoglu, M.H. Hocaoglu. Distribution systems forward/backward sweep based power flow algorithms : A review and comparison study. *Electric power component and systems, Taylor and Francis*, 3(3), 2008.

[14] Overhead Line Design Standard for Transmission & Distribution Systems, SA Power Networks, December 2012.
www.sapowernetworks.com.au/public/download.jsp?id=29686&sstat=285569

[15] J. Liu, M.M. ASalama and R.R. Mansour. an efficient power flow algorithm for distribution systems with polynomial load. *Int J. Elect.Eng. Educat.*, 39(4):371–386, 1991.

[16] J.H, Teng. +A network-topology based three phase load flow for distribution systems. *proc Natl. Sci. council ROC(A)*, 24(4):259–264, 2000.

Biographies

Souhir Mabrouk Born in Monastir, Tunisia 1978, She received her Engineering Degree in Electrical Engineering, and her DEA in automatics from National Engineering School of Tunis respectively in 2002 and 2003. She has been teaching in the High institute of technology of Sousse since 2003 and had been at the head of the security and control department from 2006 to 2008. Her interests are in the smart grid and Renewable Energy systems.

Adel Bouallegue He received his diploma in electrical engineering from National Engineering School of Tunis in 1995, He has been teaching in the High institute of technology of Sousse from 1997 to 2006. He got his PhD in electrical engineering in 2007, and he is now an assistant professor in the National Engineering School of Sousse. He is interested in sensor network, transmission, smart grid, and energy efficiency.

Adel Khedher He was born in Mahdia, Tunisia in 1976. He received his mastery of Science and DEA from ENSET of Tunis, Tunisia in 1991 and 1994, respectively, the PhD in electrical engineering from the Sfax engineering school in 2006. From 1995 to 2002, he has been a training teacher in the professional training centers. From 2003 to 2006, he has been an assistant professor in the Electronic Engineering Department of High Institute of Applied Sciences and technology. He has been promoted to the associate professor grade in the same department since March 2006. His main research interests cover several aspects related to the control of the static inverters, the Electric Machine Drives and renewable Energy Systems.

K. Koubaâ
Complex Dynamics in a Two-cell DC/DC Buck Converter using a Dynamic Feedback Controller

Abstract: In this paper, we analyze the behavior of a two-cell DC/DC buck converter controlled using a dynamic feedback controller. This converter is described by a discrete model with some simplifying hypotheses. According to the values of the parameters, we can distinguish two different cases. The theoretical conditions of stability for each case are derived using the Jury criterion. Toggling between different modes in the converter leads to the appearance of an interior crisis, sliding bifurcations and chaotic behavior. The route to sliding can be obtained via switching and grazing scenarios in the transient regime. In addition, the presence of saturation in the control law causes a serious degradation of the converter performance and shows undesirable saturating regimes. The complex dynamics and the strange phenomena encountered in the two-cell converter are confirmed with numerical simulations.

Keywords: Two-cell DC/DC buck converter, Dynamic feedback controller, Interior crisis, Sliding bifurcations, Switching and grazing scenarios, Undesirable saturating regimes, Complex dynamics.

1 Introduction

Power converters are known to exhibit a wealth of nonlinear dynamics and strange phenomena, resulting from the cyclic switching of the different topologies of these circuits. Indeed, their basic operation is essentially characterized by toggling among a set of linear or nonlinear circuit topologies. As such, they can be regarded as switched or piecewise-smooth dynamical systems [1]. Therewith, there are also several unavoidable sources of unwanted nonlinearity in practical power converters: semiconductor switching devices (transistors and diodes), nonlinear components (capacitances, inductances and power diodes) and control methods (PWM, phase-locked loops, and digital controllers) [2]. Therefore, the feedback controlled power converters routinely exhibit various types of complex dynamics, such as: chaos [3], strange attractors [4], coexistence of regular and chaotic behavior [5], [6], border collision bifurcation [7], crises [7], grazing bifurcation [8], period-doubling cascades [9], quasi-periodicity [6]–[10] and phase-locking [6], [10].

K. Koubaâ: K. Koubaâ, CEMLab, National Engineering School of Sfax, University of Sfax, Tunisia, email: karama1484@gmail.com

De Gruyter Oldenbourg, ASSD – Advances in Systems, Signals and Devices, Volume 7, 2018, pp. 61–88.
https://doi.org/10.1515/9783110470529-005

In order to reduce the consequences of switches defects and to improve the output waveforms, the multi-cell converters have been proposed firstly for high voltage applications by splitting and distributing the input voltage on intermediate levels. Recently, this class of power converters has been shown to present many other advantages over stand alone converters in term of ripple, magnetic components size and efficiency [11]–[13]. Moreover, the use of discrete time modeling [14]–[16] for the multi-cell converters combined with a digital pulse width modulation (DPWM) controller leads to discontinuous iterated maps due to the existence of various operating modes and control saturations. Then, the overall operation is similar to a piecewise smooth nonlinear dynamical system [17], which can undergo sliding bifurcations caused by the interactions between a periodic orbit and the boundary of the sliding region. Basically, four distinct cases of such bifurcations can be identified: crossing-sliding, grazing-sliding, switching-sliding and adding-sliding [18].

The existence of sliding orbits has been pointed out in the context of relay feedback systems [19], [20], dry-friction oscillator [21] and superconducting resonator [22].

Recent works focus on the occurrence of sliding bifurcations in multi-cell converters. In fact, Di Bernardo *et al.* [23] are the first to indicate the presence of a double spiralling bifurcation in the one-cell DC/DC buck converter where the unfolding of sliding leads naturally to spiralling bifurcation, which corresponds to the traditional grazing-sliding transition. In [24], a bifurcation analysis of a switched two-cell DC/DC buck converter behavior, controlled with a dynamic PI, shows the existence of the grazing-sliding in a spiralling form with multi-sliding orbits.

The aim of this work is to analyze the behavior of a two-cell DC/DC buck converter controlled using a generalized dynamic feedback controller. This converter is characterized by an interior crisis, a switching-sliding bifurcation, and especially a grazing-sliding bifurcation for a particular choice of the control parameters that leads to a dynamic PI controller. Furthermore, we notice the presence of undesirable saturating regimes caused by the saturation in the dynamic control law.

The rest of the paper is outlined as follows. Section 2 is dedicated to introduce the simplified discrete model for the two-cell DC/DC buck converter. Section 3 deals with the stability analysis and the complex dynamics in both cases of the generalized dynamic controller. Finally, concluding remarks relating the overall study and future work will be drawn in the last section.

2 Modeling of the DC/DC buck converter

The converter that we deal with in this paper is shown in Fig. 1.

It consists of two cells separated by a flying capacitor. In each cell, a switch S_j and a diode D_j ($j \in \{1, 2\}$) are turned ON and OFF in a cyclic and complementary manner, under the command of a Digital Pulse Width Modulator (DPWM). The role of the capacitor is to balance the switch voltages mainly in high voltage applications. i_L

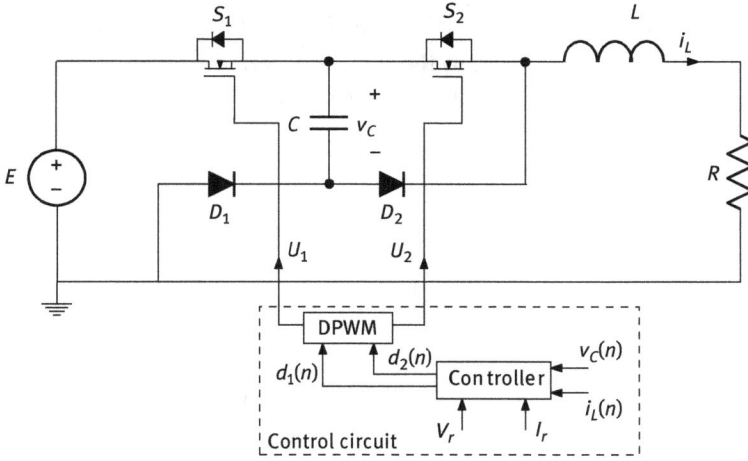

Fig. 1. Two-cell DC/DC buck converter.

and v_C are respectively the inductor current and the voltage across the capacitor. U_1 and U_2 are the outputs of the DPWM driven by a feedback control system to achieve a constant voltage $v_C = \dfrac{E}{2}$ and a constant output current $i_L = I_r$. U_1 and U_2 are phase shifted by π in order to obtain optimum waveforms of the inductor current, and represent the evolution of the duty cycles d_1 and d_2 of switches S_1 and S_2 and should vary between 0 and 1.

In this study, we will define the duty cycles with respect to the OFF state rather than the ON state. Consequently, four different topologies can be defined in Tab. 1 according to the states of the switches.

Tab. 1. Topologies in the two-cell DC/DC converter.

Topologies	State of S_1	State of S_2
Topology 1 ($\mathcal{T}_1$)	OFF	ON
Topology 2 ($\mathcal{T}_2$)	ON	ON
Topology 3 ($\mathcal{T}_3$)	ON	OFF
Topology 4 ($\mathcal{T}_4$)	OFF	OFF

To obtain the model of the two-cell DC/DC buck converter, we consider the new state variables given by:

$$x_i = \frac{R}{E} i_L \tag{1}$$

$$x_v = \frac{1}{E} v_C \tag{2}$$

where the current and the voltage are normalized by their maximum values $i_{L_{max}} = \dfrac{E}{R}$ and $v_{C_{max}} = E$ and time is also normalized by the switching period T. In each topology, the system can be described by a linear continuous model:

$$\dot{x} = A_k x + B_k \quad \text{for} \quad 1 \leq k \leq 4 \tag{3}$$

where x is the state vector:

$$x = \begin{bmatrix} x_i \\ x_v \end{bmatrix} \tag{4}$$

and matrices A_k and B_k are given by:

$$A_1 = \begin{bmatrix} -\delta_L & \delta_L \\ -\delta_C & 0 \end{bmatrix}, \ A_2 = \begin{bmatrix} -\delta_L & 0 \\ 0 & 0 \end{bmatrix}, \ A_3 = \begin{bmatrix} -\delta_L & -\delta_L \\ \delta_C & 0 \end{bmatrix}, \ A_4 = \begin{bmatrix} -\delta_L & 0 \\ 0 & 0 \end{bmatrix}$$

$$B_1 = B_4 = \begin{bmatrix} 0 \\ 0 \end{bmatrix}, \ B_2 = B_3 = \begin{bmatrix} \delta_L \\ 0 \end{bmatrix}$$

δ_L and δ_C are both the time constants:

$$\delta_L = \frac{RT}{L} \ , \ \delta_C = \frac{T}{RC} \tag{5}$$

In order to reduce the ripple current through the load and the ripple voltage across the capacitor, the circuit parameters should satisfy:

$$\delta_L \ll 1 \ \text{and} \ \delta_C \ll 1 \tag{6}$$

According to the values of the duty cycles d_1 and d_2 at the beginning of the period, toggling between different topologies occurs and leads to six possible configurations, to each of which we can express the values of the states at the beginning of the $(n+1)^{th}$ period in terms of its value at the nth period in a recurrent system. Then, we obtain the simplified discrete model [25]:

$$x_{n+1} = \mathcal{A} \, x_n + \mathcal{B} \tag{7}$$

where:

$$\mathcal{A} = \begin{bmatrix} 1 - \delta_L & \delta_L(d_1 - d_2) \\ \delta_C(d_2 - d_1) & 1 \end{bmatrix}, \mathcal{B} = \begin{bmatrix} \delta_L(1 - d_1) \\ 0 \end{bmatrix}$$

In the sequel, we choose the following values for numerical simulations:

$$\delta_L = \delta_C = 0.1 \ , \ I_r = 0.6 \ A \ , \ V_r = 0.5 \ V \ , \ E = 1 \ V$$

3 Generalized dynamic feedback controller

Delayed feedback control (DFC) is an alternative control method for chaotic systems proposed by Pyragas [26], in which the control input is fed by the difference between the current state and the delayed state. The delay time is determined as the period of the unstable periodic orbit to be stabilized [27]. However, it has a limitation such that it can not stabilize any systems with an odd number of real eigenvalues greater than one. To overcome the limitation, Yamamoto *et al.* [27] have supplemented DFC by Dynamic DFC with a function which stores the past differences as the state of the controller, and can be described in term of a difference equation of the controller's state. The proposed controller is given in the following form:

$$x_d(n+1) = A\, x_d(n) + B\, y(n) \tag{8}$$

$$u(n) = C\, x_d(n) + D\, y(n) \tag{9}$$

where $y(n) = x(n) - x(n-1)$, $A \in \mathbb{R}^{n_c \times n_c}$, $B \in \mathbb{R}^{n_c \times n}$, $C \in \mathbb{R}^{l \times n_c}$, $D \in \mathbb{R}^{l \times n}$, $x_d \in \mathbb{R}^{n_c}$ is the state of the controller, $u(n) \in \mathbb{R}^l$ is the dynamic control law and $x \in \mathbb{R}^n$ is the state [27]. Recently, this controller has been used successfully to analyze the stability of the two-cell DC/DC buck converter in [28].

3.1 Controller design

In this section, we use the same control approach without a delay at the levels of the duty cycles d_1 and d_2 defined as follows:

$$d_1(n) = \mathrm{sat}[k_i e_i(n) + k_v e_v(n) + u(n)] \tag{10}$$

$$d_2(n) = \mathrm{sat}[k_i e_i(n) + u(n)] \tag{11}$$

where k_i and k_v denote respectively the current gain and the voltage gain, $e_i(n)$ and $e_v(n)$ are the errors defined with respect to the reference values I_r and V_r as:

$$e_i(n) = x_i(n) - I_r \tag{12}$$

$$e_v(n) = x_v(n) - V_r \tag{13}$$

$u(n)$ is the output of the dynamic generalized feedback controller, and we can define the function sat by:

$$\mathrm{sat}(d) = \begin{cases} 0, & \text{if } d < 0 \\ d, & \text{if } 0 \le d \le 1 \\ 1, & \text{if } d > 1 \end{cases} \tag{14}$$

In order to guarantee a zero static error, we modify the input of the dynamic delayed controller in a feedback current error. This choice is well justified in [29], where it has been proved that we can achieve a zero static voltage error without inserting an extra state variable in the voltage loop, since its transfer function is a pure integrator. Then, the dynamic controller can be defined in the following form:

$$x_d(n+1) = a\, x_d(n) + b\, e_i(n) \tag{15}$$

$$u(n) = c\, x_d(n) + d\, e_i(n) \tag{16}$$

where a, b, c and d are scalar parameters.

Let's introduce the state vector:

$$x(n) = \begin{bmatrix} x_i(n) \\ x_v(n) \\ x_d(n) \end{bmatrix} \tag{17}$$

The closed-loop system becomes:

$$x_i(n+1) = [1 - \delta_L(1 + d + k_i)]x_i(n) - \delta_L k_v(1 + V_r)x_v(n) - \delta_L c x_d(n)$$
$$+ \delta_L k_v x_v^2(n) + \delta_L[1 + (d + k_i)I_r + k_v V_r] \tag{18}$$

$$x_v(n+1) = \delta_C k_v V_r x_i(n) + x_v(n) - \delta_C k_v x_v(n)x_i(n) \tag{19}$$

$$x_d(n+1) = b x_i(n) + a x_d(n) - b I_r \tag{20}$$

The fixed point denoted by (x_i^*, x_v^*, x_d^*) is obtained by solving the equation:

$$x^*(n+1) = x^*(n) \tag{21}$$

and is given by:

$$x_i^* = \frac{1 - a + [(1 - a)(d + k_i) + bc]I_r}{(1 - a)(1 + d + k_i) + bc} \tag{22}$$

$$x_v^* = V_r \tag{23}$$

$$x_d^* = \frac{b(1 - I_r)}{(1 - a)(1 + d + k_i) + bc} \tag{24}$$

3.2 Stability analysis

To carry out the stability analysis, we consider the error vector:

$$e(n) = x(n) - x_r \tag{25}$$

where:

$$x_r = \begin{bmatrix} I_r \\ V_r \\ \dfrac{1 - I_r}{c} \end{bmatrix} \tag{26}$$

is the reference vector.

Then, the closed-loop error system is given by:

$$e_i(n+1) = [1 - \delta_L(1 + d + k_i)]e_i(n) + \delta_L k_v(V_r - 1)e_v(n) - \delta_L c e_d(n)$$
$$+\delta_L k_v e_v^2(n) \tag{27}$$

$$e_v(n+1) = (1 - \delta_C k_v I_r)e_v(n) - \delta_C k_v e_v(n)e_i(n) \tag{28}$$

$$e_d(n+1) = b e_i(n) + a e_d(n) + (a-1)x_d^* \tag{29}$$

The characteristic equation of the linearized system is:

$$[\lambda - (1 - \delta_C k_v x_i^*)]P(\lambda) = 0 \tag{30}$$

where:

$$P(\lambda) = \lambda^2 + a_1\lambda + a_0 \tag{31}$$

with:

$$a_1 = \delta_L(1 + d + k_i) - a - 1 \tag{32}$$

$$a_0 = a[1 - \delta_L(1 + d + k_i)] + \delta_L bc \tag{33}$$

Generally, the stability condition for discrete systems is that the norm of each eigenvalue should be less than one. By applying this condition to the first eigenvalue $\lambda_1 = 1 - \delta_C k_v x_i^*$, we get:

$$0 < k_v < \frac{2}{\delta_C x_i^*} \tag{34}$$

According to the Jury criterion for the other eigenvalues, we obtain the following conditions:

$$|a_0| < 1 \tag{35}$$

$$1 + a_1 + a_0 > 0 \tag{36}$$

$$1 - a_1 + a_0 > 0 \tag{37}$$

It is easy to remark that system (15) is stable for $0 < a \leq 1$.

For $a \neq 1$, the stability conditions become:

$$-k_i + \frac{bc}{a} + \frac{1}{\delta_L}\left(1 - \frac{1}{a}\right) - 1 < d < -k_i + \frac{bc}{a} + \frac{1}{\delta_L}\left(1 + \frac{1}{a}\right) - 1 \tag{38}$$

$$d > -k_i - \frac{bc}{1-a} - 1 \tag{39}$$

$$d < -k_i + \frac{bc}{1+a} + \frac{2}{\delta_L} - 1 \tag{40}$$

From these conditions, we depict in Fig. 2 the stability zone in the plane $k_i - d$ for $0 < a < 1$.

The stability boundaries are defined by:

$$L_1 : d(k_i) = -k_i + \frac{bc}{a} + \frac{1}{\delta_L}\left(1 - \frac{1}{a}\right) - 1 \tag{41}$$

$$L_2 : d(k_i) = -k_i + \frac{bc}{a} + \frac{1}{\delta_L}\left(1 + \frac{1}{a}\right) - 1 \tag{42}$$

$$L_3 : d(k_i) = -k_i - \frac{bc}{1-a} - 1 \tag{43}$$

$$L_4 : d(k_i) = -k_i + \frac{bc}{1+a} + \frac{2}{\delta_L} - 1 \tag{44}$$

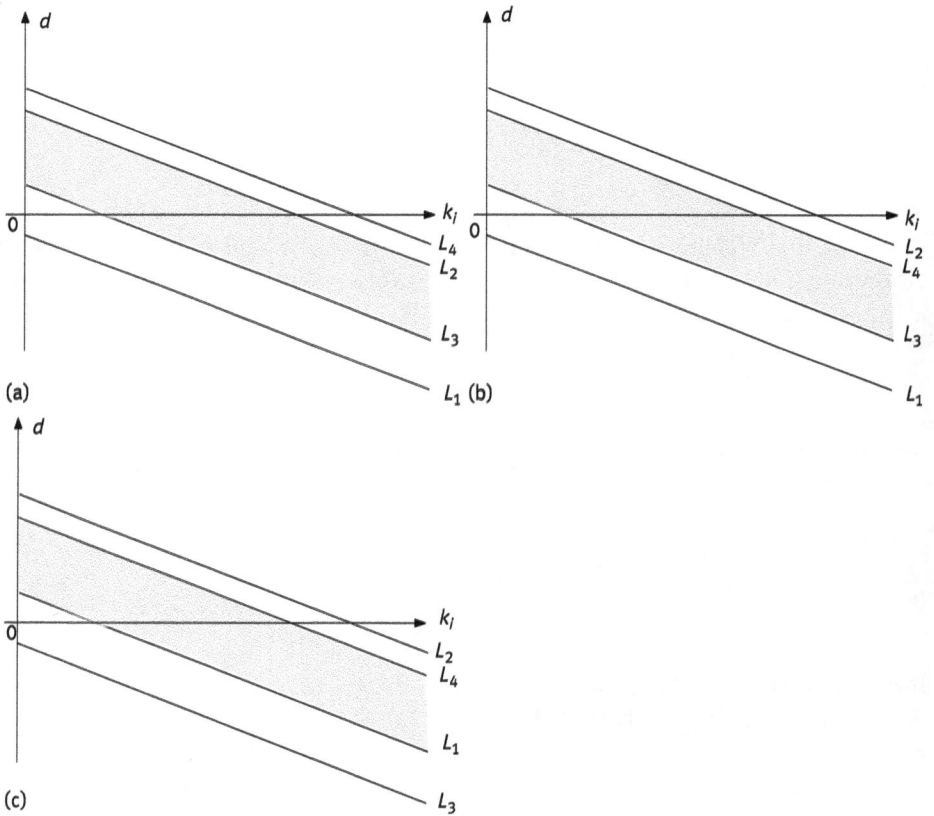

Fig. 2. Stability zone in the plane $k_i - d$ for $0 < a < 1$:
(a) $bc < \frac{1}{\delta_L}(a^2 - 1)$. (b) $\frac{1}{\delta_L}(a^2 - 1) < bc < \frac{1}{\delta_L}(1 - a)^2$. (c) $bc > \frac{1}{\delta_L}(1 - a)^2$.

The stability conditions (38)–(40) can be reformulated in the plane $b - c$ as follows:

$$\frac{-\dfrac{1+a}{\delta_L} + a(1+d+k_i)}{b} < c < \frac{\dfrac{1-a}{\delta_L} + a(1+d+k_i)}{b} \tag{45}$$

$$c > -\frac{(1-a)(1+d+k_i)}{b} \tag{46}$$

$$c > \frac{(1+a)[\delta_L(1+d+k_i) - 2]}{\delta_L b} \tag{47}$$

and are presented graphically in Fig. 3, and according to the parameters values, we can distinguish two different stability zones. The particular value $a = 1$ has different stability conditions in the plane $k_i - d$ given by:

$$-k_i + bc - 1 < d < -k_i + bc + \frac{2}{\delta_L} - 1 \tag{48}$$

$$bc > 0 \tag{49}$$

$$d < -k_i + \frac{bc}{2} + \frac{2}{\delta_L} - 1 \tag{50}$$

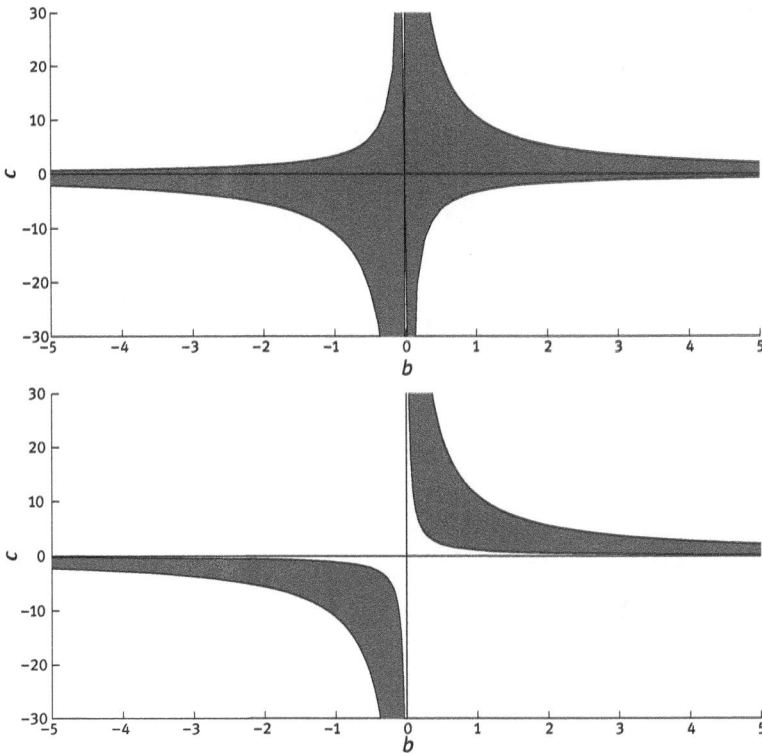

Fig. 3. Stability zones in the plane $b - c$ for $0 < a < 1$.

Then, we present the corresponding stability zone in Fig. 4, where the lines are given by:

$$L_5 : d(k_i) = -k_i + bc - 1 \tag{51}$$

$$L_6 : d(k_i) = -k_i + bc + \frac{2}{\delta_L} - 1 \tag{52}$$

$$L_7 : d(k_i) = -k_i + \frac{bc}{2} + \frac{2}{\delta_L} - 1 \tag{53}$$

In a similar way, the stability conditions (48)–(50) are reformulated in the plane $b - c$:

$$\frac{1 + d + k_i - \dfrac{2}{\delta_L}}{b} < c < \frac{1 + d + k_i}{b} \tag{54}$$

$$bc > 0 \tag{55}$$

$$c > \frac{2(1 + d + k_i) - \dfrac{4}{\delta_L}}{b} \tag{56}$$

and are graphically presented in Fig. 5, which is reduced compared to the stability zones for $0 < a < 1$ shown in Fig. 3, where C_1, C_2 and C_3 are the curves given by:

$$C_1 : c(b) = \frac{1 + d + k_i - \dfrac{2}{\delta_L}}{b} \tag{57}$$

$$C_2 : c(b) = \frac{1 + d + k_i}{b} \tag{58}$$

$$C_3 : c(b) = \frac{2(1 + d + k_i) - \dfrac{4}{\delta_L}}{b} \tag{59}$$

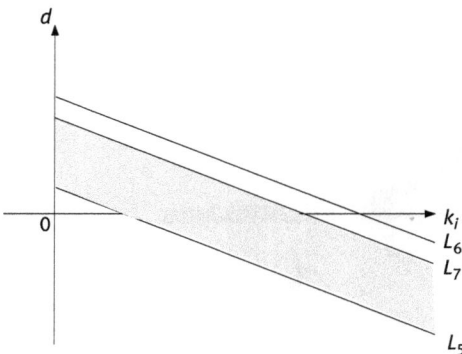

Fig. 4. Stability zone in the plane $k_i - d$ for $a = 1$.

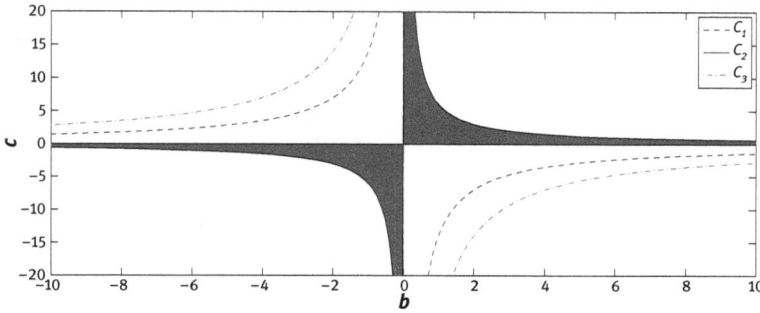

Fig. 5. Stability zone in the plane $b - c$ for $a = 1$.

3.3 Complex dynamics

We study the complex dynamics and the strange phenomena observed in the two-cell DC/DC buck converter using the Generalized Dynamic Feedback Controller. For numerical simulations, we choose the voltage gain value $k_v = \dfrac{1}{\delta_C I_r} = 16.6667$.

In Fig. 6, we present the evolution of the current $x_i(n)$ for the parameters $b = 2$, $c = 1$, $d = -15$ and for different values of the parameter a and the current gain k_i. From Figure 6 (a)–(b), we notice that when k_i increases for the same value of the parameter $a = 0.1$, the current static error diminishes.

For $k_i = 30$ and different values of the parameter a, we can reduce the current static error if we increase the value of a until 0 for $a = 1$ (see Figure 6 (b)–(c)–(d)). Figure 7 and 8 show the existence of undesirable saturating regimes in the current response corresponding to $x_i(n) = 1$ and $x_i(n) = 0$ due to the saturation of the duty cycles respectively at 0 and 1.

Figure 9 presents the 2D bifurcation diagrams for the current behavior in the plane $k_i - d$ for the parameters $b = 2$ and $c = 1$ and for three values of the parameter a. Here, we use different colors to indicate the periodic behaviors from period 1 to period 20 and two undesirable saturating regimes that correspond to the current $x_i(n) = 0$ and $x_i(n) = 1$ respectively in light blue and red. Brown refers to the 1 periodic behavior and black is used for chaos.

The 2D bifurcation diagrams confirm the stability zone obtained theoretically. We can easily remark the existence of a five periodic behavior presented in green for the case $a = 1$.

From Figures 7 and 8, we can determine analytically the expressions of the boundaries of saturation of the duty cycles as follows:

$$x_i(n) = 1 \implies d_2(n) = 0 \implies d = -k_i - \frac{bc}{1 - a}$$

$$x_i(n) = 0 \implies d_1(n) = 1 \implies d = -k_i - \frac{bc}{1 - a} - \frac{1 + k_v V_r}{I_r}$$

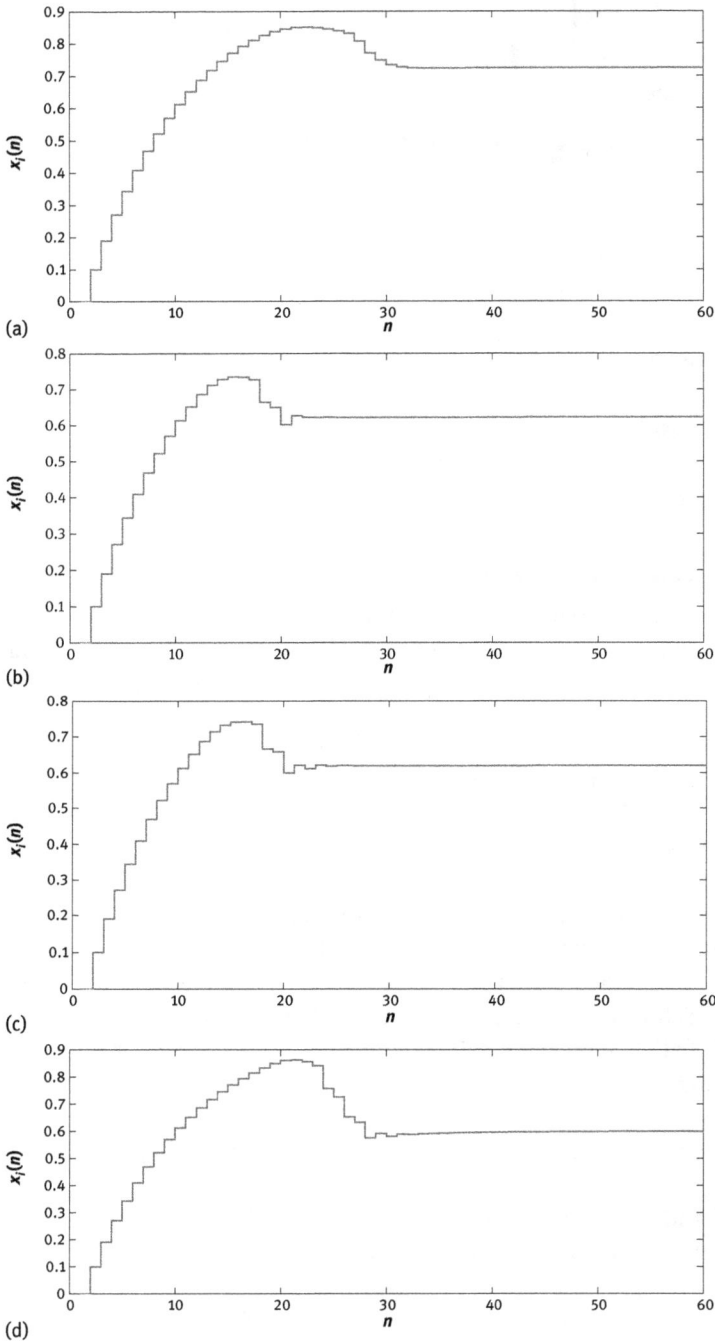

Fig. 6. Evolution of the current $x_i(n)$ for the parameters $b = 2$, $c = 1$, $d = -15$. (a) $a = 0.1$ and $k_i = 15$. (b) $a = 0.1$ and $k_i = 30$. (c) $a = 0.5$ and $k_i = 30$. (d) $a = 1$ and $k_i = 30$.

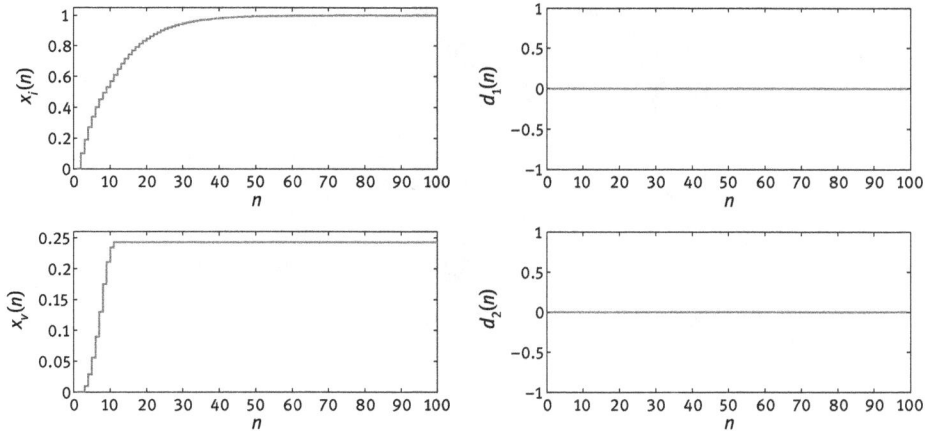

Fig. 7. Current response for the parameters: $a = 0.1$, $b = 2$, $c = 1$, $k_i = 10$ and $d = -20$.

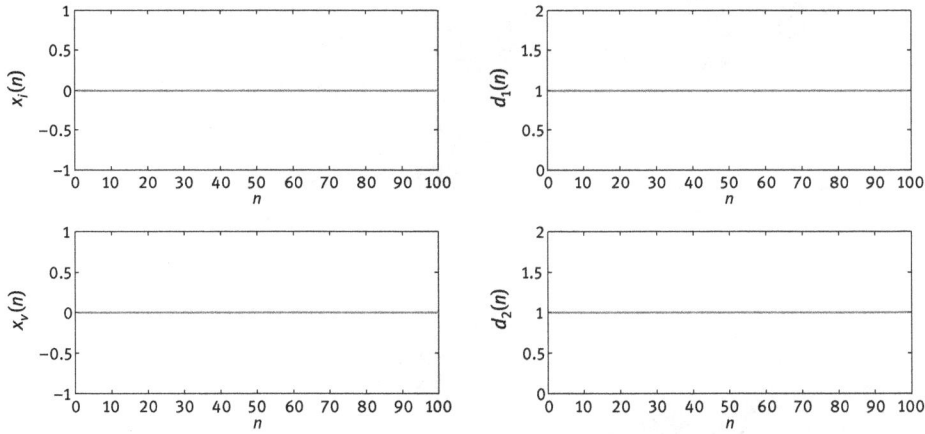

Fig. 8. Current response for the parameters: $a = 0.1$, $b = 2$, $c = 1$, $k_i = 10$ and $d = -30$.

Figure 10 presents the 2D bifurcation diagrams of the current behavior in the plane $b - c$ for different values of the parameters, which confirms the two cases of the stability zone depicted in Fig. 3. In Figure 11, we plot the 2D bifurcation diagrams of the current behavior in the plane $b - c$ for the parameters $k_i = 40$ and $d = -25$ and for the three values of the parameter $a = 0.1, 0.5, 1$. The stable zone of the one periodic behavior is represented in brown. In addition to the saturating regimes and the five periodic behavior, we notice the presence of the period 2, the period 4 and high periodic orbits.

In Fig. 12, we present the 1D bifurcation diagrams for parameters chosen from the 2D bifurcation diagrams in the plane $k_i - d$ and $d = -25$.

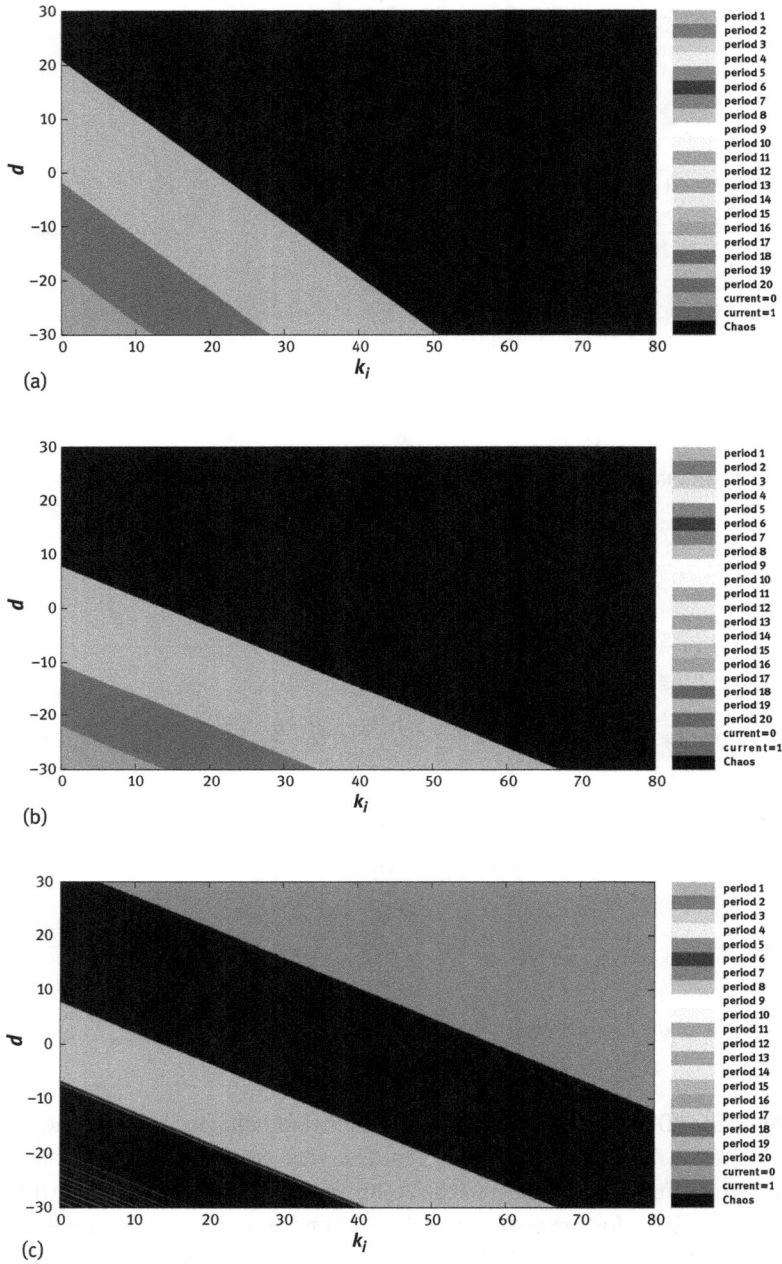

Fig. 9. 2D bifurcation diagrams of the current behavior in the plane $k_i - d$ for the parameters $b = 2$ and $c = 1$: (a) $a = 0.1$. (b) $a = 0.5$. (c) $a = 1$.

(a)

(b)

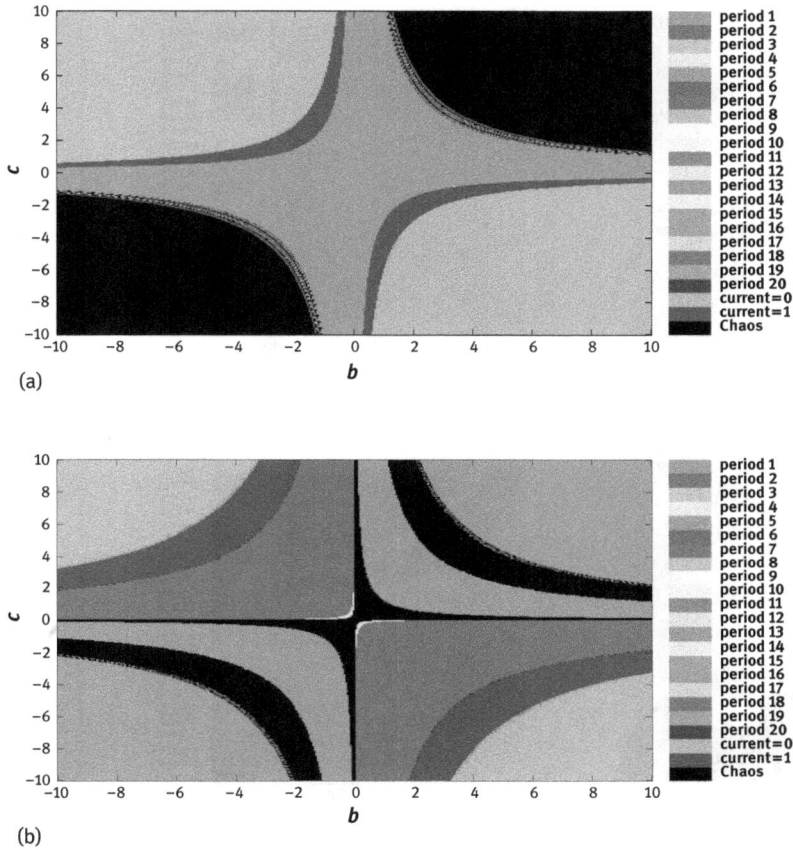

Fig. 10. 2D bifurcation diagrams of the current behavior in the plane $b - c$ for the parameters:
(a) $a = 0.7$, $k_l = 40$ and $d = -30$. (b) $a = 0.1$, $k_l = 40$ and $d = -20$.

For $a = 0.1$ and $a = 0.5$, the current behavior is characterized by two saturating regimes $x_i(n) = 0$ and $x_i(n) = 1$, then a one periodic behavior with a high static error, which bifurcates in a two periodic behavior at one bifurcation point in a degenerate flip bifurcation, and after we obtain a chaotic behavior followed by a five-piece attractor. The third 1D bifurcation diagram for $a = 1$ exhibits other complex dynamics. We can distinguish four main parts in this diagram. Indeed, for low values of the current gain $k_i < 26.05$, high periodic windows are alternated with quasi-periodic strips, which characterizes the phase-locking phenomenon. One distinctive feature is that the two-cell DC/DC buck converter under a delayed feedback controller [6], [28] exhibits low periodic behaviors in the phase-locking region. It has been reported in [30] that the delayed feedback controller is applicable to the stabilization of quasi-periodic orbits. Then, a one periodic behavior appears after a subcritical

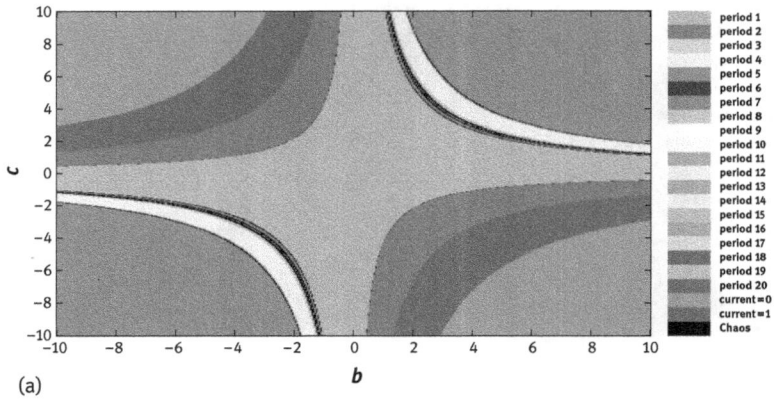

(a)

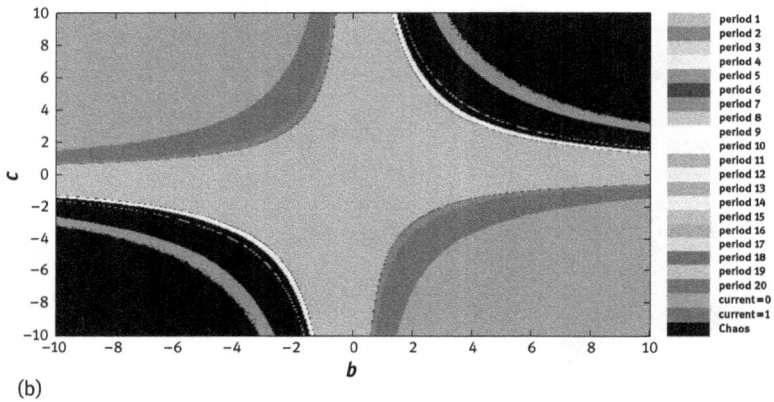

(b)

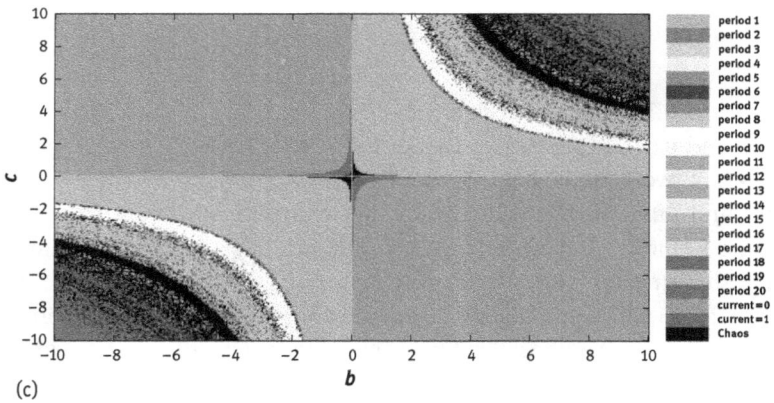

(c)

Fig. 11. 2D bifurcation diagrams of the current behavior in the plane $b - c$ for the parameters $k_l = 40$ and $d = -25$: (a) $a = 0.1$. (b) $a = 0.5$. (c) $a = 1$.

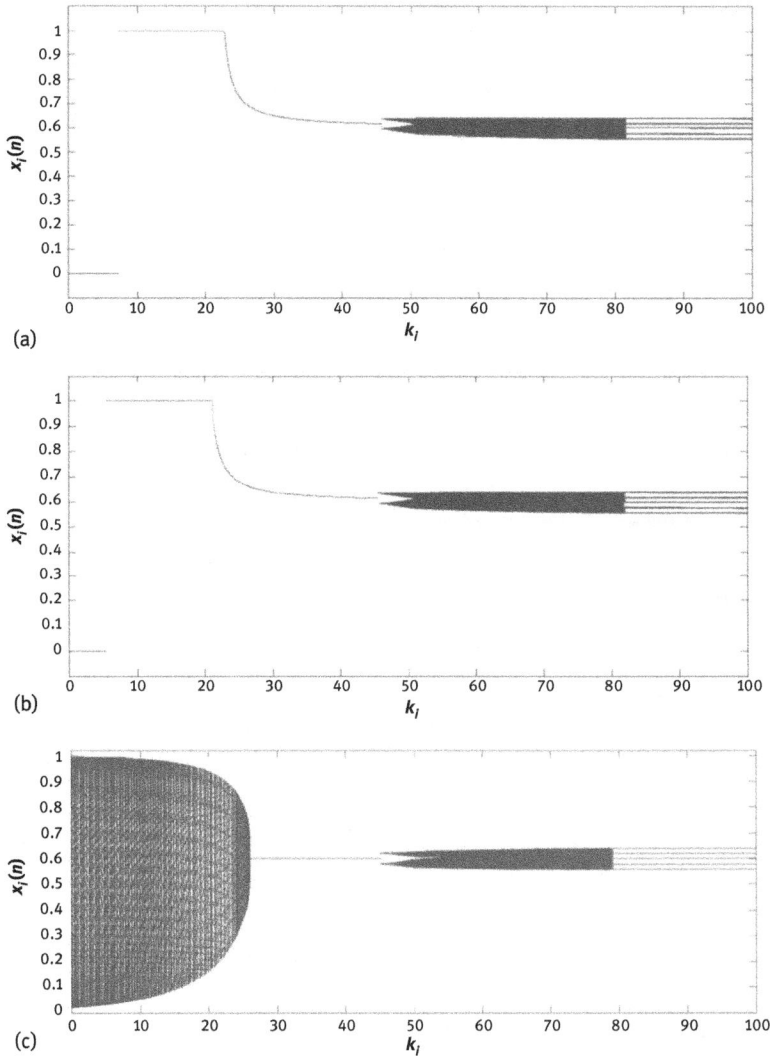

Fig. 12. 1D bifurcation diagrams of the current $x_I(n)$ versus k_i for the parameters $b = 2$, $c = 1$ and $d = -25$: (a) $a = 0.1$. (b) $a = 0.5$. (c) $a = 1$.

Neimark-Sacker bifurcation based on two complex conjugate eigenvalues crossing the unit circle. We obtain a two periodic behavior at one bifurcation point $k_i = 45$ followed by a chaotic behavior in a degenerate flip bifurcation, and a five periodic behavior for high values of the current gain $k_i > 79.09$. We carry out a zoom in the previous 1D bifurcation diagrams in Fig. 13.

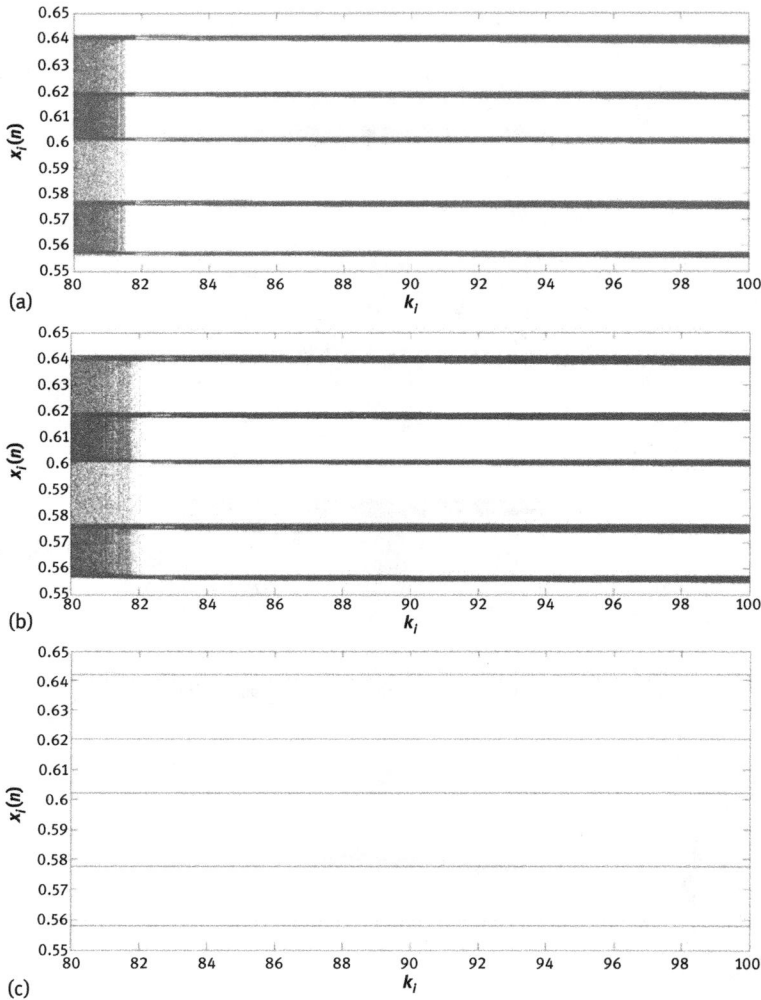

Fig. 13. Zoom in the 1D bifurcation diagrams of the current $x_i(n)$ versus k_i for the parameters $b = 2$, $c = 1$ and $d = -25$: (a) $a = 0.1$. (b) $a = 0.5$. (c) $a = 1$.

We remark for $a = 0.1, 0.5$ the appearance of a five-piece attractor suddenly from a one chaotic attractor. It has been reported in [31] that almost all sudden changes in the size of chaotic attractors are due to crises. It is an interior crisis, which is characterized by the sudden expansion/contraction of the size of a chaotic attractor [32], and dues to the collision of the chaotic attractor with the unstable periodic orbit in its basin of attraction [33]. In the case $a = 1$, we obtain a five periodic behavior.

In the sequel, we will focus on the types of bifurcations that occur in the two-cell DC/DC buck converter. Therefore, we plot the phase spaces for the parameters $b = 2, c = 1$, and $d = -25$ for $a = 0.1, 0.5, 1$ in Figures 14, 16 and 18 at the steady regime for the initial condition $x_v(1) = 0.5$. Figure 14 and 16 illustrate the interior crisis for $0 < a < 1$, and to obtain more explanation, we represent a zoom in the phase space in Figure 15 and 17. For $a = 0.1$, before crisis, at $k_i = 81.51$, we have a chaotic attractor, but after crisis, at $k_i = 81.5101$, the strange attractor becomes five-piece, and similarly for $a = 0.5$. Figure 18 shows the phase space $x_d - x_i$ for two different values of the current gain $k_{i_1} = 70$ and $k_{i_2} = 90$ that indicate respectively the presence of chaos and five periodic behavior in the 1D bifurcation diagram presented in Fig. 12 (c).

We plot the lower and upper boundaries of saturation of the duty cycles at 0 and 1 respectively in light blue and red and their expressions are given by:

$$d_1(n) = d_2(n) = 0 \Longleftrightarrow x_i(n) = -\frac{c}{k_i + d} x_d(n) + I_r$$

$$d_1(n) = d_2(n) = 1 \Longleftrightarrow x_i(n) = -\frac{c}{k_i + d} x_d(n) + \frac{1}{k_i + d} + I_r$$

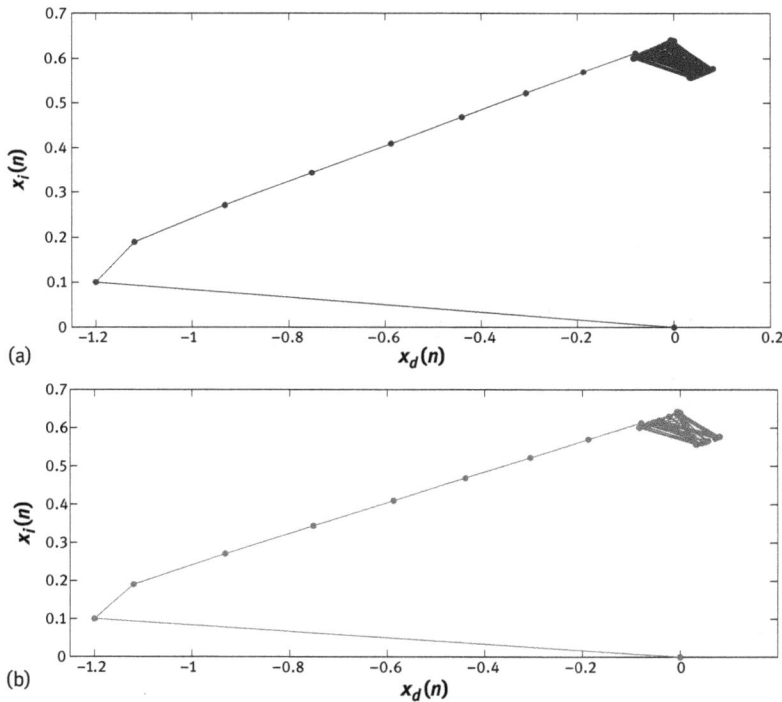

(a)

(b)

Fig. 14. Phase space $x_d - x_i$ for the parameters $a = 0.1$, $b = 2$, $c = 1$, and $d = -25$. (a) before crisis: $k_i = 81.51$. (b) after crisis: $k_i = 81.5101$.

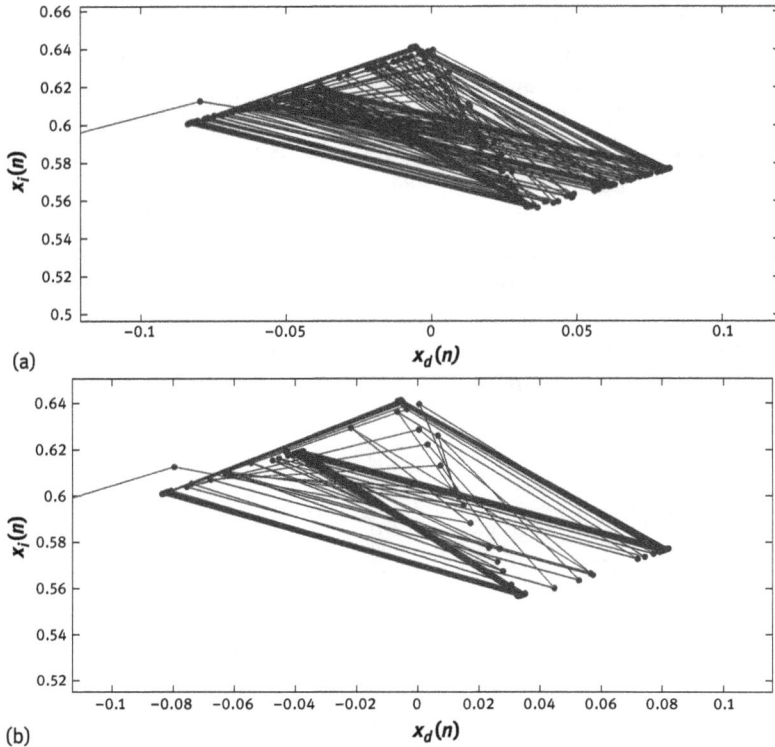

Fig. 15. Zoom in the phase space $x_d - x_i$ for the parameters $a = 0.1$, $b = 2$, $c = 1$, and $d = -25$: (a) before crisis: $k_l = 81.51$. (b) after crisis: $k_l = 81.5101$.

The saturation boundaries divide the phase space into three regions: region of non saturation delimited by the two saturation lines, region of saturation at 0 under the blue line and region of saturation at 1 above the red line. The current behavior is characterized by successive switchings between the three regions and it slides into five periodic orbit as it is shown in Fig. 19 (b).

In order to have sliding motion in the discrete case, the necessary and sufficient condition is:

$$|S(n+1)| < |S(n)| \tag{60}$$

where $S(n)$ is the switching function [34].

The sliding orbit is obtained for the non saturated value of the duty cycle $d_1(94) = -2.5598$ and the next value $d_1(95) = 0.1791$ which verifies condition (60):

$$|d_1(95)| < |d_1(94)| \tag{61}$$

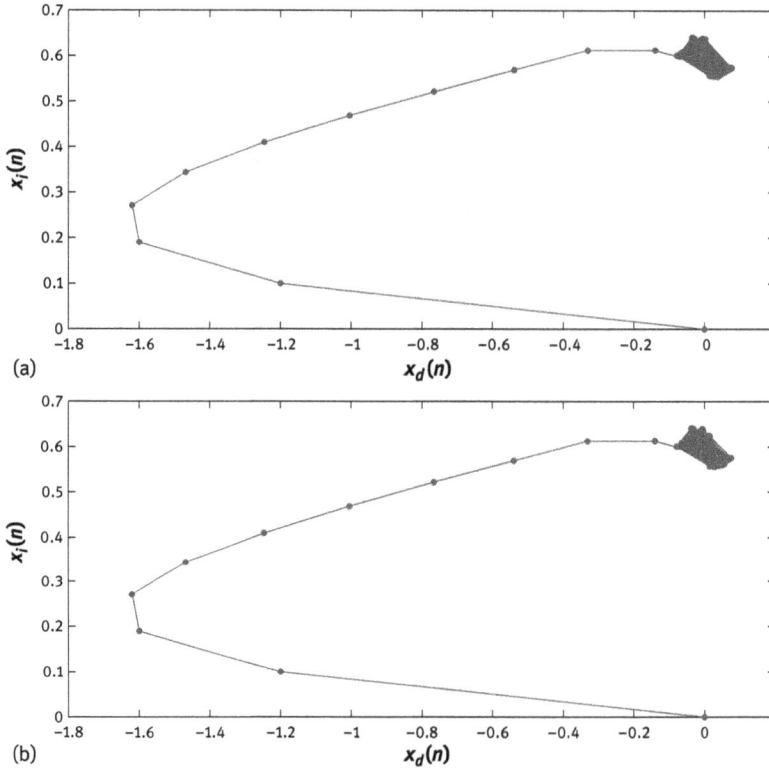

Fig. 16. Phase space $x_d - x_i$ for the parameters $a = 0.5$, $b = 2$, $c = 1$, and $d = -25$. (a) before crisis: $k_i = 81.8$. (b) after crisis: $k_i = 81.81$.

In Figure 20, we present the evolution of the current $x_i(n)$ and the duty cycles $d_1(n)$ and $d_2(n)$ for the parameters $a = 1$, $b = 2$, $c = 1$, $k_i = 90$ and $d = -25$, that show a five periodic behavior and the duty cycles characterized by a successive switching between the non saturation, saturation at 0 and saturation at 1.

In [24], for a particular choice of the parameters $a = c = 1$ and $b = d = \dfrac{k_i}{\tau_i}$ that leads to a dynamic PI controller, we obtain a grazing-sliding bifurcation where periodic orbits characterized by different numbers of switchings accumulate onto a sliding orbit.

Figure 21 shows the 1D bifurcation diagram of the current $x_i(n)$ versus b for the parameters $a = 1$, $c = 6$, $d = -25$, and $k_i = 40$. For negative values of the parameter b, we obtain a saturating regime ($x_i(n) = 0$), which is confirmed in the 2D bifurcation diagram. For $b = 0$ and according to equations (15)–(16), the dynamic feedback

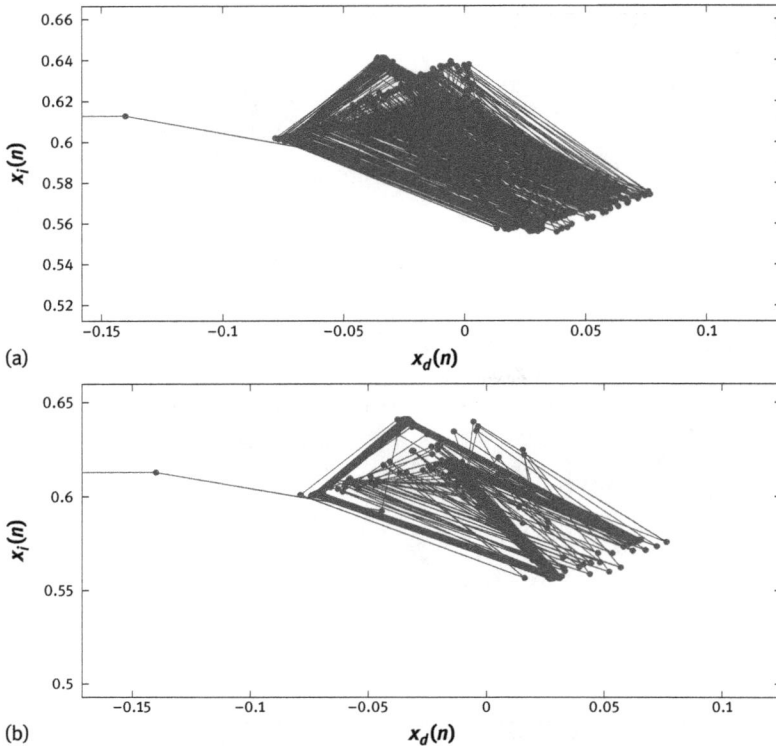

Fig. 17. Zoom in the phase space $x_d - x_i$ for the parameters $a = 0.5$, $b = 2$, $c = 1$, and $d = -25$:
(a) before crisis: $k_i = 81.8$. (b) after crisis: $k_i = 81.81$.

controller behaves as a proportional controller. For positive values of the parameter b, three behaviors can be shown:
- one periodic behavior for $0 < b < 2.6$.
- phase-locking phenomenon for $2.6 \leq b \leq 2.85$.
- high periodic orbits for $b > 2.85$.

The one periodic behavior is illustrated with the example treated in Fig. 22, which proves the presence of the windup phenomenon characterized by excessive oscillations, large overshoot and long settling time caused by the successive switching of the duty cycles at 0 then at 1.

Figure 23 shows the phase space $x_d - x_i$ for high periodic orbits. We remark the existence of a spiralling bifurcation also called grazing-sliding bifurcation. The global behavior consists of a transient regime and a steady regime. The transient regime shows grazing bifurcation points at the corners of the spiral after successive saturation phases of the duty cycles at 0 then at 1 (see Fig. 24), and the steady regime is a sliding

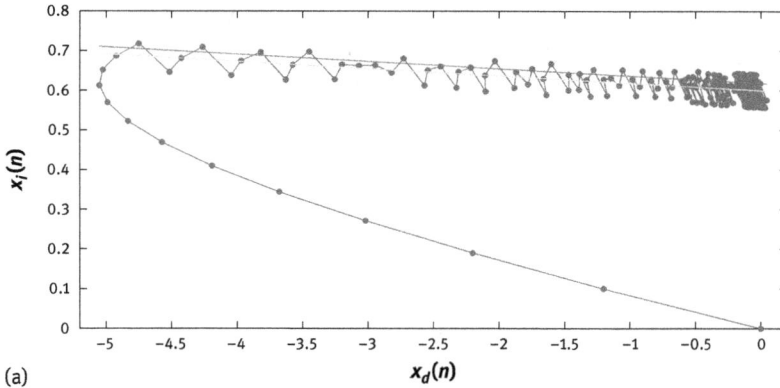

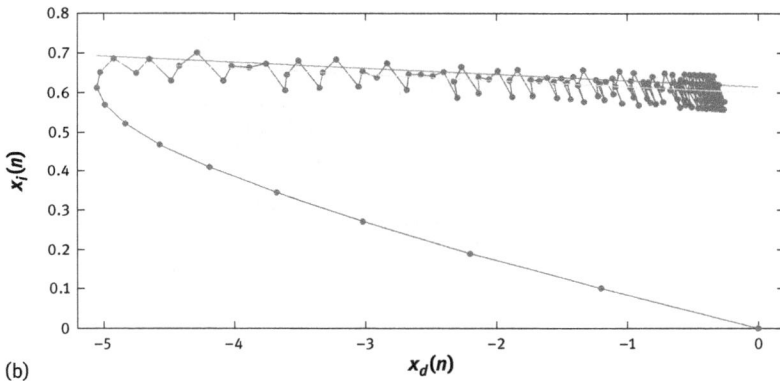

Fig. 18. Phase space $x_d - x_i$ for the parameters $a = 1, b = 2, c = 1$, and $d = -25$. (a) $k_{i_1} = 70$: Chaotic behavior. (b) $k_{i_2} = 90$: Five periodic behavior.

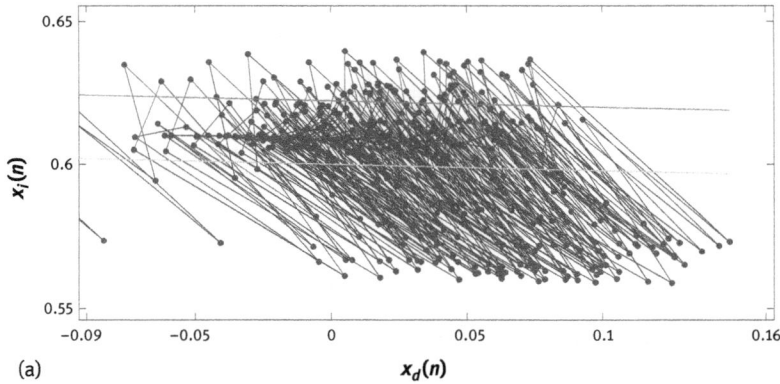

Fig. 19. Zoom in the phase space $x_d - x_i$ for the parameters $a = 1, b = 2, c = 1$, and $d = -25$. (a) $k_{i_1} = 70$: Chaotic behavior. (b) $k_{i_2} = 90$: Five periodic behavior.

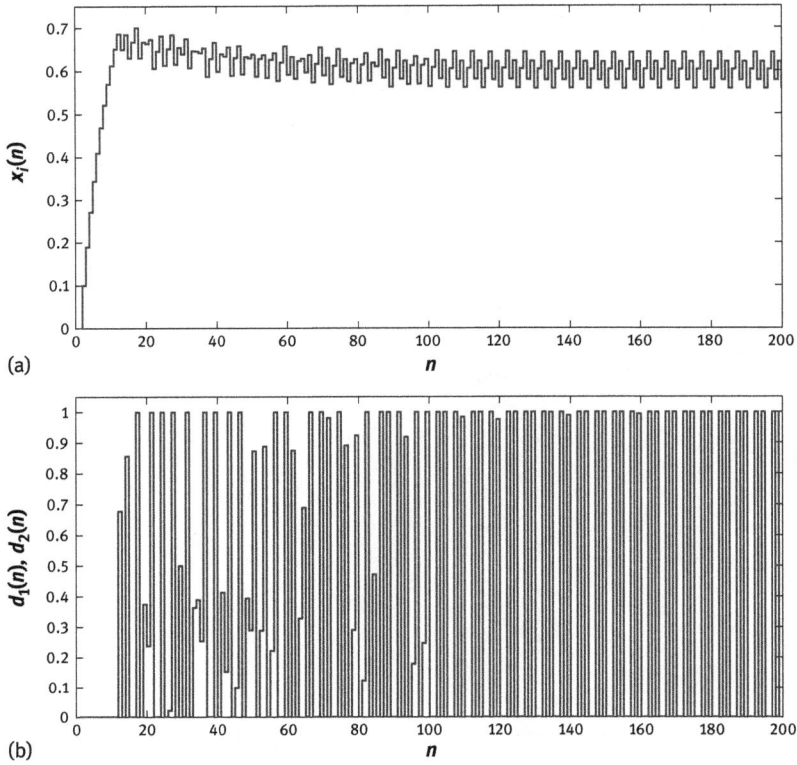

Fig. 20. Evolution of the current and the duty cycles for the parameters $a = 1$, $b = 2$, $c = 1$, $k_l = 90$ and $d = -25$: (a) current $x_i(n)$. (b) duty cycles $d_1(n)$, $d_2(n)$.

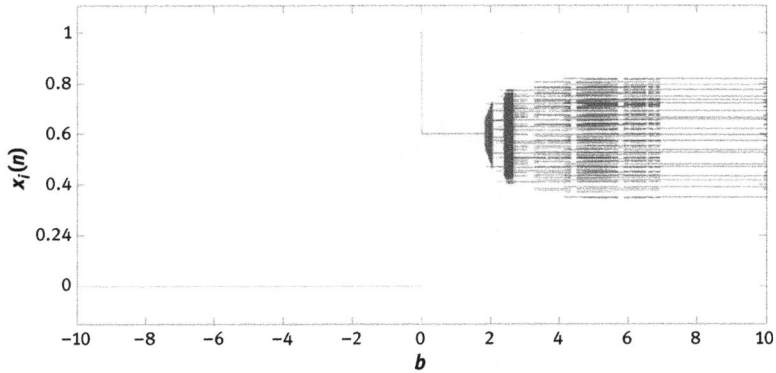

Fig. 21. 1D bifurcation diagram of the current $x_i(n)$ versus b for the parameters $a = 1$, $c = 6$, $d = -25$, and $k_l = 40$.

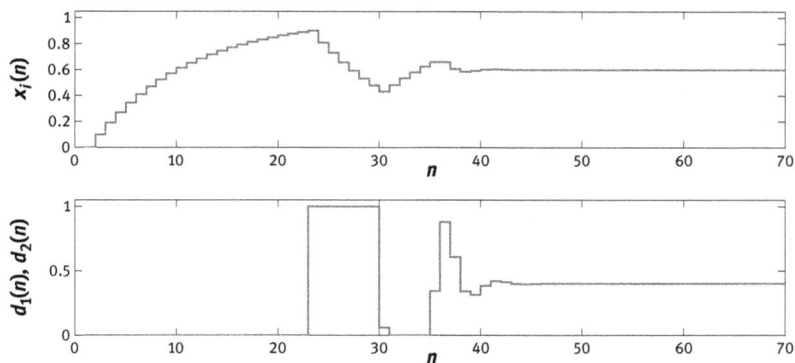

Fig. 22. Evolution of the current for the parameters $a = 1$, $b = 1.5$, $c = 6$, $k_i = 40$, and $d = -25$.

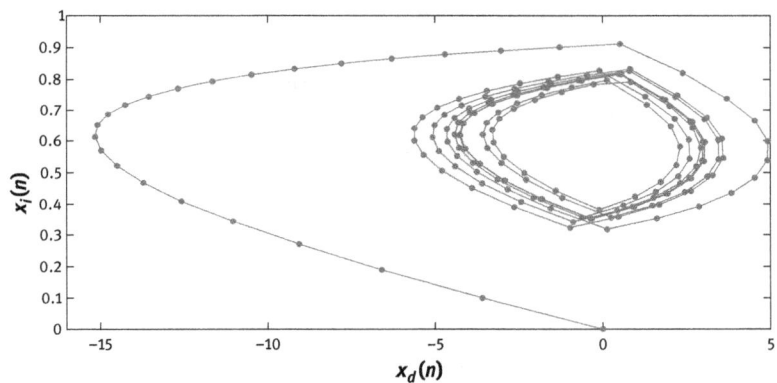

Fig. 23. Phase space $x_d - x_i$ for the parameters $a = 1$, $b = 6$, $c = 10$, $k_i = 40$, and $d = -25$.

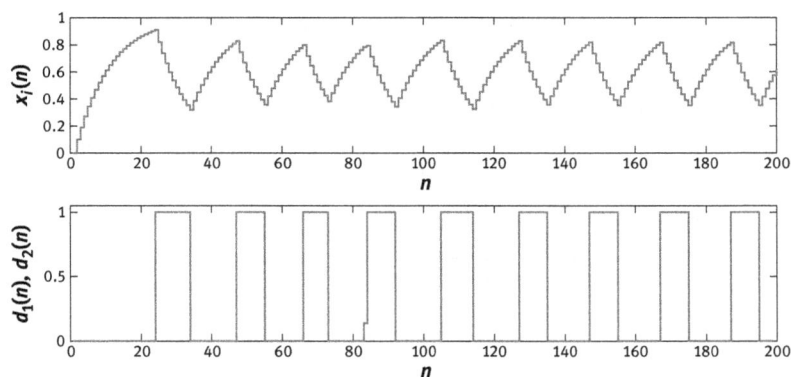

Fig. 24. Evolution of the current and the duty cycles for the parameters $a = 1$, $b = 6$, $c = 10$, $k_i = 40$, and $d = -25$.

motion of a high periodic orbit in the middle of the spiral. This type of bifurcation has been analyzed and the grazing boundaries are given analytically for a dynamic PI controller in a previous work [24], and the condition for sliding to a high periodic orbit is given by (60).

4 Conclusion

In this work, we have investigated the complex dynamics in the two-cell DC/DC buck converter under a dynamic feedback controller. Interior crisis, switching-sliding and grazing-sliding bifurcations are present in the current behavior of the converter, and conditions to obtain a sliding motion have been derived in the discrete case according to the values of the duty cycles. In addition, it has been demonstrated the existence of undesirable saturating regimes with constant values of the current at 0 and 1 when the duty cycles saturate. The two-cell DC/DC buck converter exhibits also the phase-locking phenomenon for high periodic orbits, the Neimark-Sacker bifurcation and the degenerate flip bifurcation. In a future work, we will analyze the complex dynamics in other types of power converters such as multi-level and stacked multi-cell converters.

Bibliography

[1]   C.K. Tse and M. di Bernardo. Complex Behavior in Switching Power Converters. *Proceedings of the IEEE*, 90(5):768–781, 2002.
[2]   D.C. Hamill, S. Banerjee and G.C. Verghese. *Nonlinear Phenomena in Power Electronics: Bifurcations, Chaos Control, and Applications*. Wiley-IEEE Press, July. 2001.
[3]   M. di Bernardo and C.K. Tse. *Chaos in Power Electronics: An Overview, Chaos in Circuits and Systems*. Chapter 16, :317–340, New York: World Scientific, June. 2002.
[4]   M-C. Péra, B. Robert and C. Goeldel. Non linear Dynamic in Electromechanical Systems-Application to a Hybrid Stepping Motor. *Electromotion*, 7:31–42, 2000.
[5]   K. Koubaâ, M. Feki, A. El Aroudi and B.G.M. Robert. Coexistence of Regular and Chaotic Behavior in The Time-Delayed Feedback Controlled Two-Cell DC/DC Converter. *Int. Conf. on Systems, Signals & Devices*, Djerba-Tunisia, March 23–26, 2009.
[6]   K. Koubaâ and M. Feki. Quasi-Periodicity, Chaos and Coexistence in the Time Delay Controlled Two-cell DC–DC Buck Converter. *Int. Journal of Bifurcation and Chaos*, 24(10):1450124, 2014.
[7]   G. Yuan, S. Banerjee, E. Ott and J.A. Yorke. Border-Collision Bifurcations in the Buck Converter. *IEEE Trans. on Circuits and Systems–I: Fundamental Theory and Applications*, 45(7):707–716, 1998.
[8]   M. di Bernardo, E. Fossas, G. Olivar and F. Vasca. Secondary Bifurcations and High Periodic Orbits in Voltage Controlled Buck Converter. *Int. Journal of Bifurcation and Chaos*, 7(12):2755–2771, 1997.
[9]   C.K. Tse. Flip Bifurcation and Chaos in a Three-state Boost Switching Regulator. *IEEE Trans. on Circuits and Systems–I: Fundamental Theory and Applications*, 41(1):16–23, 1994.

[10] A. El Aroudi and R. Leyva. Quasi-Periodic Route to Chaos in a PWM Voltage-Controlled DC–DC Boost Converter. *IEEE Trans. on Circuits and Systems–I: Fundamental Theory and Applications*, 48(8):967–978, 2001.

[11] A. El Aroudi, J. Peláez, M. Feki and B.G.M. Robert. Stability Analysis of Two-cell Buck Converter Driven DC Motor with a Discrete-time Closed Loop. In *Int. Conf. on Systems, Signals & Devices*, Djerba-Tunisia, March 23–26, 2009.

[12] T.A. Meynard, H. Foch, P. Thomas, J. Courault, R. Jakob and M. Nahrstaedt. Multicell Converters: Basic Concepts and Industry Applications. *IEEE Trans. on Industrial Electronics*, 49(5):955–964, 2002.

[13] T.A. Meynard, M. Fadel and N. Aouda. Modeling of Multilevel Converters. *IEEE Trans. on Industrial Electronics*, 44(3):356–364, 1997.

[14] A. El Aroudi, B. Robert and L. Martínez-Salamero. Modelling and Analysis of Multi-Cell Converters Using Discrete Time Models. In *the IEEE Int. Symp. on Circuits and Systems*, :2161–2164, Island of Kos-Greece, May 21–24, 2006.

[15] B. Robert and A. El Aroudi. Discrete Time Model of a Multi-cell dc/dc Converter: Non Linear Approach. *Mathematics and Computers in Simulation*, 71(4 & 6):310–319, 2006.

[16] A. El Aroudi, B.G.M. Robert, A. Cid-Pastor and L. Martínez-Salamero. Modeling and Design Rules of a Two-Cell Buck Converter Under a Digital PWM Controller. *IEEE Trans. on Power Electronics*, 23(2):859–870, 2008.

[17] B. Robert, M. Feki and H.H.C. Iu. Control of a PWM Inverter Using Proportional Plus Extended Time-Delayed Feedback. *Int. Journal of Bifurcation and Chaos*, 16(1):113–128, 2006.

[18] I. Merillas Santos. *Modeling and Numerical Study of Nonsmooth Dynamical Systems. Applications to Mechanical and Power Electronic Systems*. PhD thesis, Technical University of Catalonia, Spain, 2006.

[19] M. di Bernardo, K. Johansson and F. Vasca. Sliding Orbits and Their Bifurcations in Relay Feedback Systems. In *the IEEE Conf. on Decision & Control*, Phoenix, Arizona-USA, 1:708–713, December 7–10, 1999.

[20] M. di Bernardo, K.H. Johansson and F. Vasca. Self-Oscillations and Sliding in Relay Feedback Systems: Symmetry and Bifurcations. *Int. Journal of Bifurcation and Chaos*, 11(4):1121–1140, 2001.

[21] M. Guardia, S.J. Hogan and T.M. Seara. An Analytical Approach to Codimension-2 Sliding Bifurcations in the Dry-Friction Oscillator. *SIAM Journal on Applied Dynamical Systems*, 9(3):769–798, 2010.

[22] M.R. Jeffrey, A.R. Champneys, M. di Bernardo and S.W. Shaw. Catastrophic Sliding Bifurcations and Onset of Oscillations in a Superconducting Resonator. *Physical Review E*, 81(1):016213, 2010.

[23] M. di Bernardo, C. Budd and A. Champneys. Grazing, Skipping and Sliding: Analysis of the Non-smooth Dynamics of the DC/DC Buck Converter. *Nonlinearity*, 11(4):858–890, 1998.

[24] K. Koubaâ and M. Feki. Bifurcation Analysis and Anti-windup Approach for a Dynamic PI Controller in a Switched Two-cell DC/DC Buck Converter. *the Annual Seminar on Automation, Industrial Electronics and Instrumentation*, Tanger-Maroc, 25–27 June 2014.

[25] M. Feki, A. El Aroudi and B.G.M. Robert. *Multicell dc/dc Converter: Modeling, Analysis and Control*. Internal report of a Tunisian-Spanish PCI cooperation project No A/6828/06, 2007.

[26] K. Pyragas. Continuous Control of Chaos by Self-Controlling Feedback. *Physics Letters A*, 170(6):421–428, 1992.

[27] S. Yamamoto, T. Hino and T. Ushio. Dynamic Delayed Feedback Controllers for Chaotic Discrete-Time Systems. *IEEE Trans. on Circuits and Systems–I: Fundamental Theory and Applications*, 48(6):785–789, 2001.

[28] K. Koubaâ, M. Feki, A. El Aroudi, B.G.M. Robert and N. Derbel. Stability Analysis of a Two-cell DC/DC Converter Using a Dynamic Time Delayed Feedback Controller. *Int. Conf. on Systems, Signals & Devices*, Amman-Jordan, June 27–30, 2010.

[29] K. Koubaâ, J. Pelaez-Restrepo, M. Feki, B.G.M. Robert and A. El Aroudi. Improved Static and Dynamic Performances of a Two-cell DC-DC Buck Converter Using a Digital Dynamic Time-Delayed Control. *Int. Journal of Circuit Theory and Applications*, 40(4):395–407, 2012.

[30] N. Ichinose and M. Komuro. Delayed Feedback Control and Phase Reduction of Unstable Quasi-Periodic Orbits. *Chaos*, 24(3):033137, 2014.

[31] C. Grebogi, E. Ott and J.A. Yorke. Crises, Sudden Changes in Chaotic Attractors, and Transient Chaos. *Physica D: Nonlinear Phenomena*, 7(1–3):181–200, 1983.

[32] F.A. Borotto, A.C.-L. Chian and E.L. Rempel. Alfvén Interior Crisis. *Int. Journal of Bifurcation and Chaos*, 14(7):2375–2380, 2004.

[33] J. Feng and W. Xu. Chaotic Boundary Crisis in the Duffing Van der Pol Vibro-Imapct Oscillator. In *Int. Conf. on Multimedia Technology*, pp. 2431–2434, Hangzhou-China, July 26–28, 2011.

[34] S.Z. Sarpturk, Y. Istefanopulos and O. Kaynak. On the Stability of Discrete-time Sliding Mode Control Systems. *IEEE Trans. on Automatic Control*, 32(10):930–932, 1987.

Biography

Karama Koubaâ received her degree in Electrical Engineering in 2008 and then the Master in Automation and Control in 2009 from the National Engineering School of Sfax. She obtained her PhD on the control of power electrical systems and the analysis of their chaotic behavior from the University of Sfax in 2013. Recently, she is appointed as Assistant Professor in Automatic control and Industrial computing within the Higher Institute of Computer and Multimedia of Gabès. Currently, she is a postdoctoral researcher in the Control and Energy Management Laboratory of the Engineering School of Sfax. Her research interests include chaos control and bifurcation analysis mainly for multi-cell converters and power electronic circuits.

R. Yahmadi, K. Brik and F. Ben Ammar

Degradation Analysis of the Lead Acid Battery Plates in the Manufacturing Process Based on SADT and Causal Tree Analysis

Abstract: During the lead acid battery manufacturing, the chemical reactions taking place on both positive and negative plates have a significant impact on their performance and lifetime. This paper explores a degradation analysis of the lead acid battery plate during the manufacturing process. In this context, a functional study of the different manufacturing processes of lead acid battery plate with Structured Analysis and Design Technique (SADT) is presented. Then, all the causes and potential factors generating a low quality of the plate are identified by the Ishikawa diagram. Finally, the details of the degradation of lead acid battery plate are described by the Causal Tree Analysis (CTA) during the manufacturing of lead oxide, paste, grid and the processes of pasting, curing and drying.

Keywords: lead acid battery; degradation; plate; SADT; Ishikawa; causal tree.

1 Introduction

The lead acid battery is the most widely energy storage systems used and its use in different applications presents a problem related to the limit of its lifetime. However, an optimization of the battery lifetime passes through an improvement during its manufacturing and operating processes [1]. The lead acid battery is composed of the positive and negative plates immersed in an electrolyte solution. The quality of the plate during the manufacturing process has a significant impact on the lead acid battery performance. In this context, the chemical reactions vital during the manufacturing of positive and negative plate must be understood and controlled. In fact, the chemical composition, the mechanical behavior and the metallurgical parameters during the manufacturing process of the plate exhibited a profound impact on the composition and the crystal structure of the plate [2–4]. The main objective of this study is to analysis the degradation of the lead acid battery plate during the manufacturing process by adopting a reliability analysis. The reliability analysis of the lead acid battery plates is based on two stages.

R. Yahmadi, K. Brik and F. Ben Ammar: R.Yahmadi, National High School of Engineer of Tunis, Tunisia, email: yahmadiraja@gmail.com, K.Brik, Higher Institute of Multimedia Art of Manouba, Tunisia, email: kais.brik@yahoo.fr, F. Ben Ammar,National Institute of Applied Sciences and Technology, Tunisia, email: faouzi.benamar@yahoo.fr

De Gruyter Oldenbourg, ASSD – Advances in Systems, Signals and Devices, Volume 7, 2018, pp. 89–104.
https://doi.org/10.1515/9783110470529-006

The first stage is interested in the description and modeling of the various manufacturing processes of the lead acid battery plates by standard Structured Analysis and Design Technique (SADT) mythology (NF X50–150) . The different manufacturing processes of lead acid battery plate are the manufacturing of lead oxide, paste, grid, and the processes of pasting, curing and drying of the plate.

The second stage consists of degradation analysis of the lead acid battery plate during the manufacturing process by Ichikawa diagram and causal tree analysis. The Ishikawa diagram presents the relationships between the causes and the effect possible causing a low quality of the plate. However, the deductive analysis based on the causal tree presents the various combinations of events that involve the degradation of lead acid battery plates.This degradation of lead acid battery plates is generated by the qualities of lead oxide, paste, grids, pasting or curing and drying of the plate.

2 Modeling of the manufacuring process of the lead acid battery plates

In the manufacturing process of the lead acid battery plate, the lead ingots are oxidized in the presence of the airstream to produce the lead oxide. Then, the lead oxide is added to water, sulfuric acid and a range of proprietary additives in the mixers to form a paste consistency. However, the grid is manufactured by perforating of molted lead alloys in the perforating machine. During the pasting process, the paste is suitably spread onto the grids to form uncured plates. These plates are then subjected to the curing and drying processes to improve the adhesion of the paste on the grid [5, 6]. The diagram of the lead acid battery plate manufacturing is shown in Fig. 1.

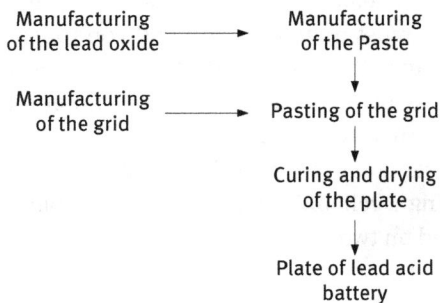

Fig. 1. Diagram of lead acid battery plate manufacturing.

2.1 Manufacturing of lead oxide

In the manufacturing of lead oxide, the soft lead (99.97 % pure) is converted into lead oxide composed of lead monoxide (PbO) and metallic lead (free lead) [5]. The lead oxides are produced by the Barton pot or the ball mill process. In this paper, the ball mill process is described and its modeling is presented in Fig. 2.

The soft lead ingots are conveyed into a rotating ball mill with a rotation speed about 25–35 rpm and a temperature range of 70–80° C [2, 7]. The lead pieces grind against each other to form smaller pieces. The appropriate size particles are oxidized to form a lead oxide and transported by airstream into classifying, cyclones and then into the bag filter [3]. After that, the fine particles of lead oxide are collected in hoppers while the coarse particles are collected and returned to rotating a ball mill. The ball mill process offers a very high content of free lead 25–35 wt % relative to 75–65 wt % of lead monoxide composed polymorphic forms "α–PbO".

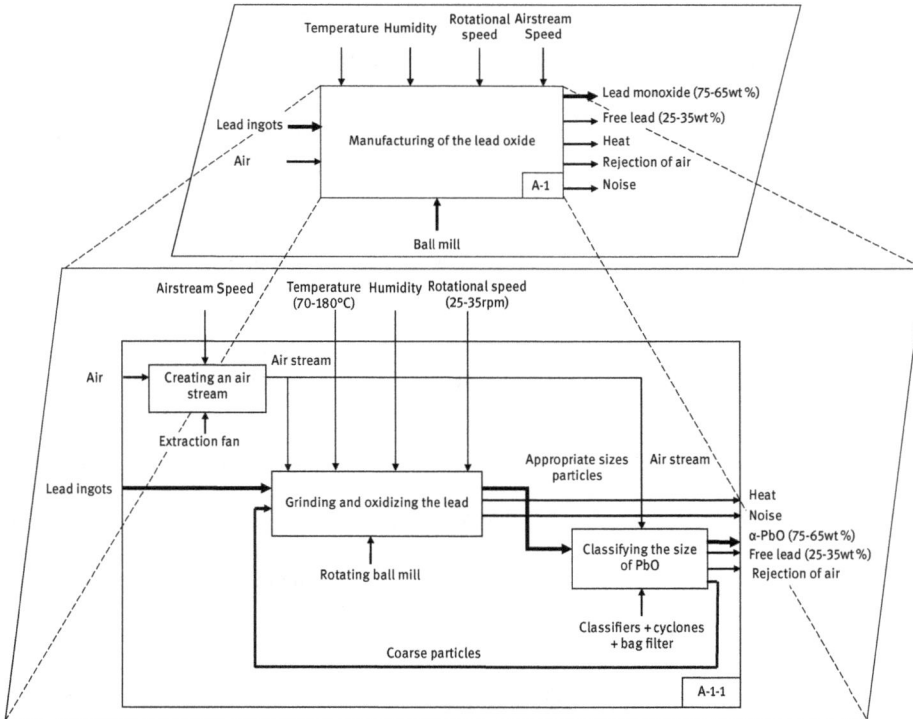

Fig. 2. Diagram of lead oxide manufacturing by Ball mill process.

2.2 Manufacturing of paste

The manufacturing process of the paste is very important as battery performance and lifetime are determined by the paste properties. The paste is prepared by mixing the lead oxide, deionized water (pH = 7) and sulfuric acid solution (density $1.40 \, cm^3$) with the rotation speed between 9–11 rpm [2]. The mixing process allows the oxidation of lead, the formation of basic lead sulfates and the formation of hydrates. The basic lead sulfates $PbSO_4$ can be produced as a mono-basic form (1 BS: $PbO.PbSO_4$), tri-basic form (3 BS: $3 \, PbO.PbSO_4.H_2O$) or tetra-basic form (4 BS: $4 \, PbO.PbSO_4$) [6].

The improvement of the paste quality such as porosity and conductivity is ensured by the addition of a range of suitable additives during the paste manufacturing. The additives added are different for the positive and negative pastes. The additives of the positive paste are designed for handling the material deposited on the positive electrode while the additives of the negative paste are considered for holding the sponge lead on the negative electrode. The pastes of positive and negative plates are prepared in the separate mixers so as to avoid the contamination of the positive paste with the additives from the negative paste that may affect their performance. Table 1 shown the various additives of the positive and negative paste.

An example of the preparation of 3 BS paste for 500 kg weight of lead oxide is shown in this study. For this weight of lead oxide, the addition of 65 L of water, 39 L

Tab. 1. Additives of the positive and negative paste [8–10]

	Additives	Operating mode
Positive paste	Diatomite(3–5 wt %)	Maintain the porosity during the restructuring of the active mass
	Glass microspheres (1.1–6.6 wt %) Expanded graphite (0.5 wt %) Barium plumbates (0.1–0.3 wt %)	Increase the porosity of the active mass
	Acetylene Black (0.2–2 wt %)	Renfort the conduction system of the active mass
	Glass microfibers (0.5–1.5 wt %)	Increase the specific surface of the active mass
	Fiber (1–2 wt %)	Stabilization of the active mass in its restructuring
Negative paste	Lignosulfonate (0.2–0.3 wt %)	Increase the surface area of the negative active mass
	Black carbon (0.1–0.3 wt %)	Increase the conductivity of the negative active mass
	Expanded graphite (<1.5 wt %)	
	Polyaspartate (<0.05 wt %)	Facilitate the formation of small crystals of lead sulfate($PbSO4$)
	Barium sulfate (0.8–1 wt %)	

of sulfuric acid solution and 0.35 kg of additives is ensured with a temperature below 50°C [2]. The modeling of the paste manufacturing is shown in Fig. 3.

The production of the 3 BS paste begins by adding of additives and water respectively into the mixer, followed by mixing them for $t_1 = 1$ minute. Then, the lead oxide is added and mixed with the previous components for $t_2 = 3-4$ minutes until a homogeneous mass is obtained [2]. The first reaction that takes place is the hydration of PbO:

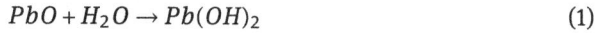

$$PbO + H_2O \rightarrow Pb(OH)_2 \tag{1}$$

Later, the sulfuric acid is added gradually for 10–12 minutes to produce the 3 BS crystals. The reaction of basic lead sulfate is described by (2).

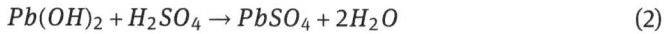

$$Pb(OH)_2 + H_2SO_4 \rightarrow PbSO_4 + 2H_2O \tag{2}$$

The reaction of monobasic lead sulfate is determined by (3).

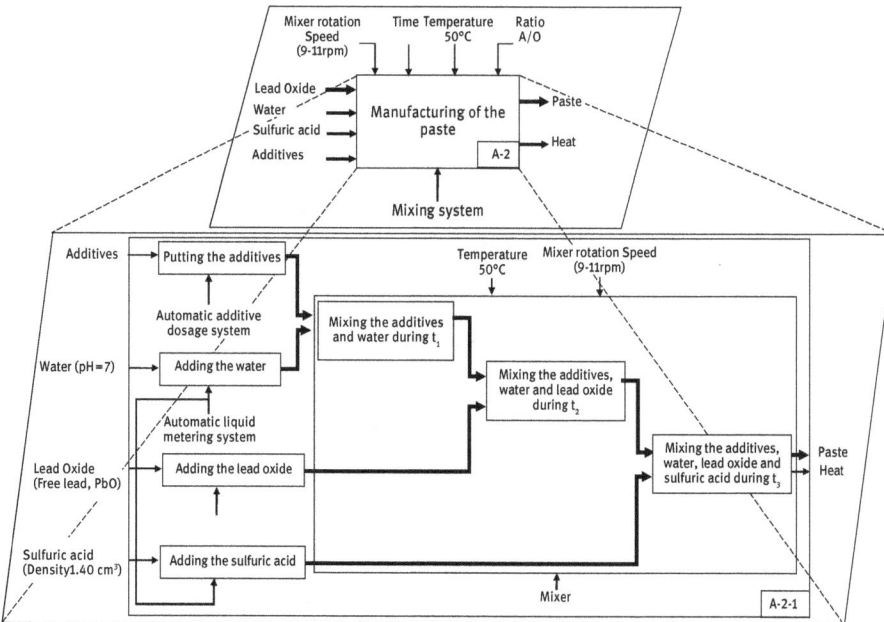

$$PbSO_4 + Pb(OH)_2 \rightarrow PbO.PbSO_4 + H_2O \tag{3}$$

Fig. 3. Diagram of the manufacturing of paste.

After 10 min of agitation of the paste, the formation of 3 BS begins at the detriment of reduced quantities of 1 BS and PbO.

$$PbO.PbSO_4 + 2Pb(OH)_2 \rightarrow 3PbO.PbSO_4.H_2O + H_2O \tag{4}$$

Finally, the obtained paste is mixed for t_3 = 15 minutes in order to allow the growth of 3 BS crystals. The hydration reaction of 3 BS is exposed by (5):

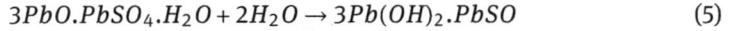

$$3PbO.PbSO_4.H_2O + 2H_2O \rightarrow 3Pb(OH)_2.PbSO \tag{5}$$

2.3 Manufacturing of grids

The lifetime and performance of the lead acid battery plate are also influenced by the quality of the grid. In fact, the grid quality depends on its composition, its design and its manufacturing process. The grid is obtained either by rolling (gravity casting technology) or perforation (expanded metal technology). In this paper, the perforation technology is described and modeled in Fig. 4.

In the perforation technology, the lead alloys are melted in an oven and then cooled to form a strip of lead. After all, the lead strip goes through the conveyor belt into the perforating machine in order to perforate in desired shape.

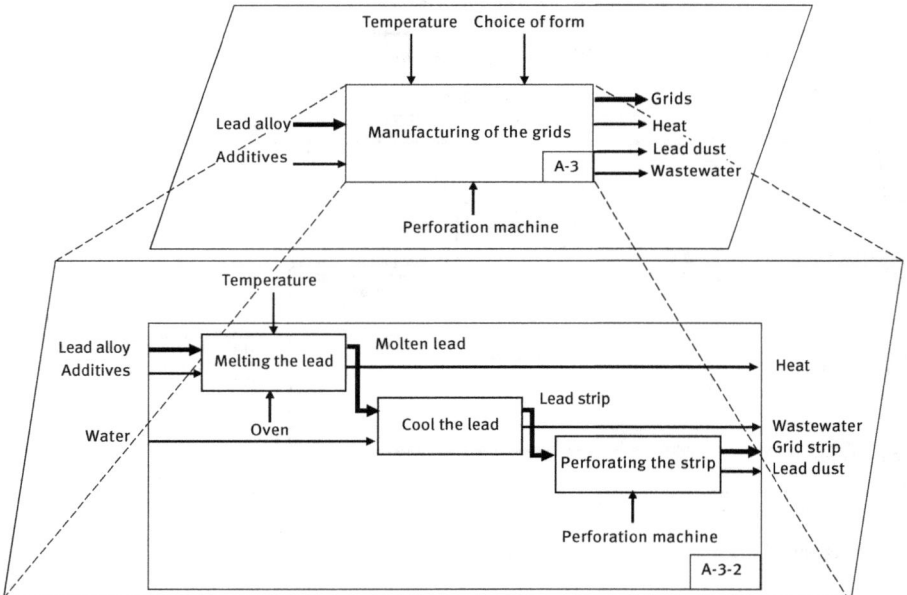

Fig. 4. Diagram of the grids manufacturing by perforation technology.

The composition of the grid must not only consider mechanical strength, creep strength, and corrosion resistance but also provides better conductivity between the grid and the active mass. The improvement of the grid composition is provided by adding additives in the lead alloy such as tin, arsenic, silver, aluminum, etc. The selection of appropriate additives during the grid manufacturing is important in order to improve the battery lifetime. The additives dosage of the grid alloy is illustrated in the Table 2.

Tab. 2. Additives of the grid alloy [11, 12].

	Additives	Operating mode
Lead-antimony	Tin (0.05–1.2 wt %)	Increase the castability.
		Reduce the rate of corrosion.
		Improve the conductivity.
	Arsenic (0.05–0.5 wt %)	Increase the creep resistance.
		Improve the corrosion resistance.
	Selenium(0.015–0.025 wt %)	Refine the grain size.
		Improve the castability.
Lead-calcium	Tin (0.3–1.2 wt %)	Better mechanical properties.
		Improve the corrosion behavior.
		No deterioration with aging.
		Improving the creep resistance.
	Barium (>0.015 wt %)	Improve both the mechanical properties and corrosion behavior.
		Prevent aging.
		Improve the creep resistance.
	Silver (0.015–0.05 wt %)	Increase the mechanical properties.
		Increase the creep resistance and the resistance to growth.
		Improve the corrosion behavior.
	Aluminum(0.008–0.012 wt %)	Avoid the preferential oxidation of calcium in the molten state.
		Prevent the loss of Calcium (Ca) during casting grid.

2.4 Pasting process

During the pasting process, the grid is first transported by a conveyor belt and placed on the pasting paper. At the same time, the paste is placed into a hopper in the pasting machine and discharged periodically on the grid. Then, the paste is pressed mechanically in different openings of the grid and the excess paste is removed using scrapers and brushes. Finally, the plate is covered with another paper to prevent the sticking of the plates. Figure 5 shows the modeling of the pasting process.

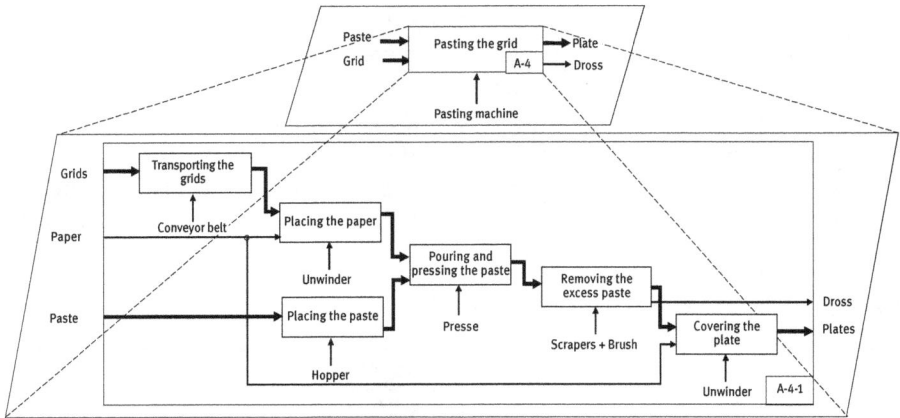

Fig. 5. Diagram of the pasting process.

2.5 Curing and drying of the plate

The curing and drying processes enable the suitable adhesion of the paste on the grid. In fact, the paste particles are connected into a solid porous mass during the curing process. Then, the free lead particles are oxidized to lead oxide and the basic lead sulfates are converted. This process ensures also the oxidation of the grid alloy and the formation of a Corrosion Layer (CL) on the grid surface to improve the adhesion between the cured paste and the grid [2]. The curing process is performed by controlling the temperature, time, airstream and humidity in a curing chamber. After the curing process, the plates are subjected to the drying process in order to improve the texture quality of the plate by a proper control of temperature and humidity. The improvements of the texture quality are more effective oxidation of free lead, better cohesion and less cracks. The modeling of curing and drying processes is shown in Fig. 6.

The performance parameters of the battery (capacity and lifetime) are strongly predetermined by the paste type used in the plate manufacturing. A high quality of plate is generally obtained from 4 BS positive pastes cured that have a high output power performance with longer cycle life than that of the plate with 3 BS positive pastes cured [4]. The conditions of curing and drying processes of the 3 BS positive paste are described in Table 3.

In the negative plate, the additives added during the manufacturing of paste don't allow the formation of 4 BS crystals. In fact, the curing and drying of the negative paste 3 BS allows the growth of large crystals 3 BS and the dissolution of smaller ones under the condition described in Table 4.

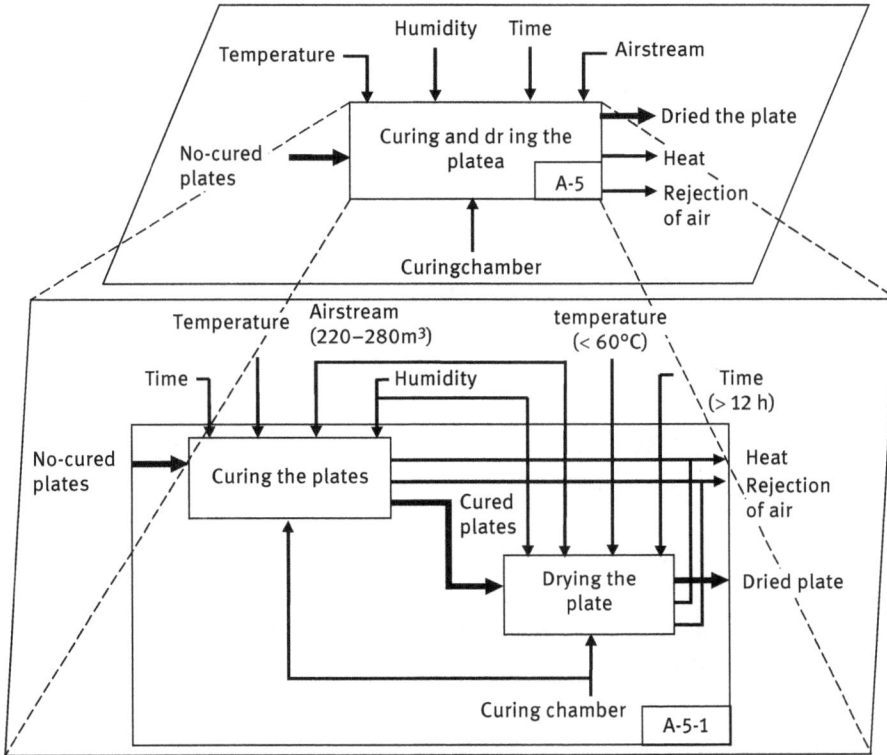

Fig. 6. Diagram of curing and drying process.

Tab. 3. Curing and drying processes of the positive paste 3 BS.

Process	Step	Condition
Curing	Recrystallization and interconnection of the particles	Temperature of 80–95° C; Relative humidity 100 %; Time 24 h.
	Oxidation of the grid and the formation of a corrosion layer	Temperature of 50° C; Relative humidity 60 %; Time 12 h.
Drying	Increase the adhesion of the cured paste on the grid	Temperature of 60° C; Relative humidity 1.0 %; Time 12 h.

Tab. 4. Curing and drying processes of the negative paste 3 BS.

Process	Step	Condition
Curing	Recrystallization and interconnection of the particles	Temperature of 45° C; Relative humidity of 95 % ; Time 12 h.
	Oxidation of the grid and the formation of a corrosion layer	Temperature of 50° C ; Relative humidity of 60 %; Time 12 h.
Drying	Increase the adhesion of the cured paste on the grid	Temperature of 60° C ; Relative humidity of 1.0 % ; Time 12 h.

3 Analysis of the lead acid battery plate degradation during the manufacturing process

The various manufacturing processes of lead acid battery plate are studied and modeled by SADT. This modeling is completed by a qualitative analysis based on the Ishikawa diagram and the causal tree in order to identify the various factors that involve the degradation of the lead acid battery plate during the manufacturing process.

3.1 Ishikawa diagram

The study of the various manufacturing processes of lead acid battery plate allows building the Ishikawa diagram. As shown in Fig. 7, Ichikawa diagram allows to identify all the causes and potential factors that cause the low quality of the plate using a structured approach. This diagram consists of six categories (people, environment, methods, materials; measurements and machines) which have different levels of precision. The Ishikawa diagram allows identifying areas where data should be collected for further study by the causal tree.

3.2 Causal trees analysis

The causal tree analysis (CTA) is a powerful technique recognized by an international standard object (IEC61025). This method allow to identify all possible combinations of causes that involves the low quality of the plate using a tree structure. In fact, each event is generated from the events of the lower level via various logical operators [13]. As shown by Fig. 8, the low quality of the plate is caused by: Fault quality of pasting, curing or drying processes.

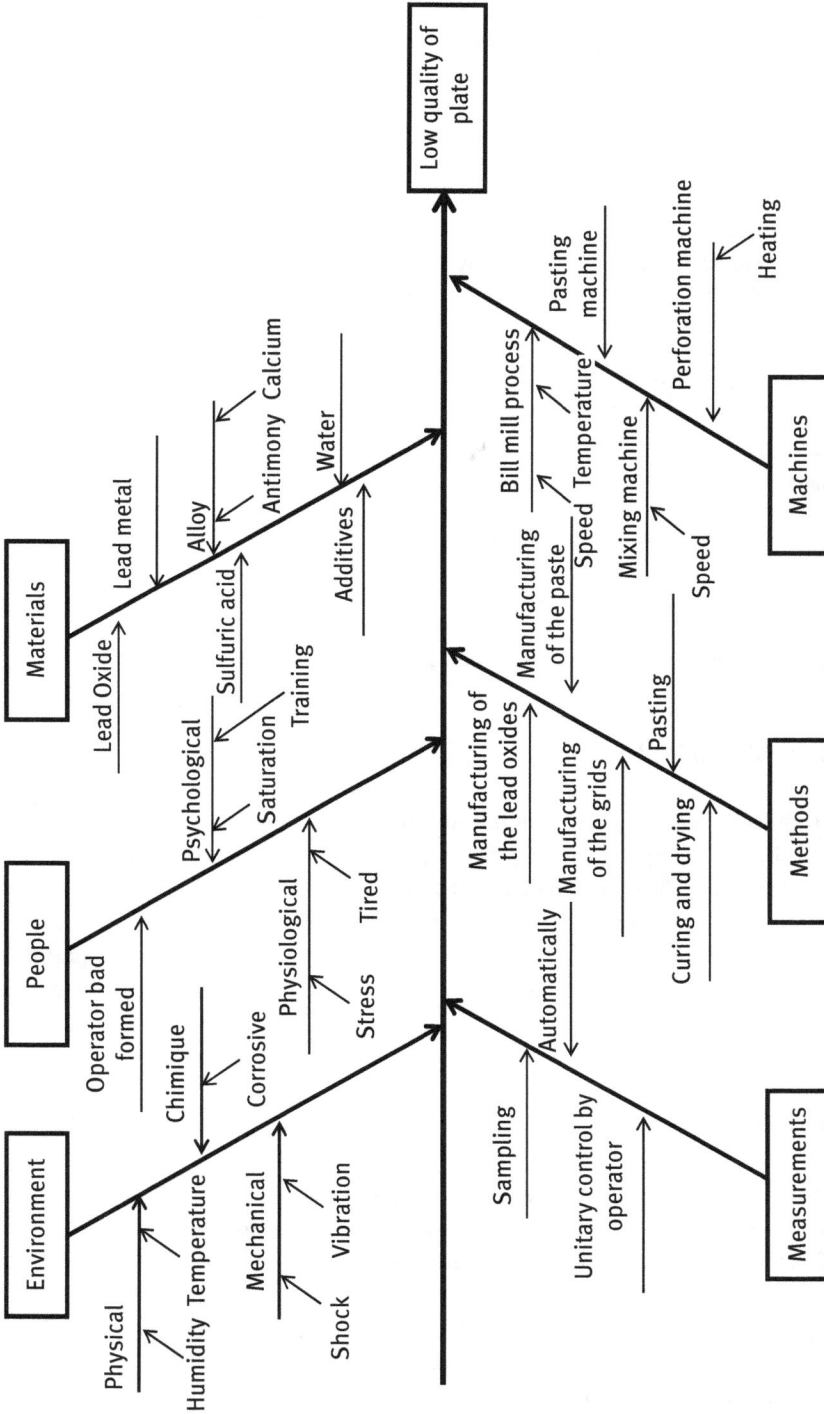

Fig. 7. Ishikawa diagram of the low quality of plate.

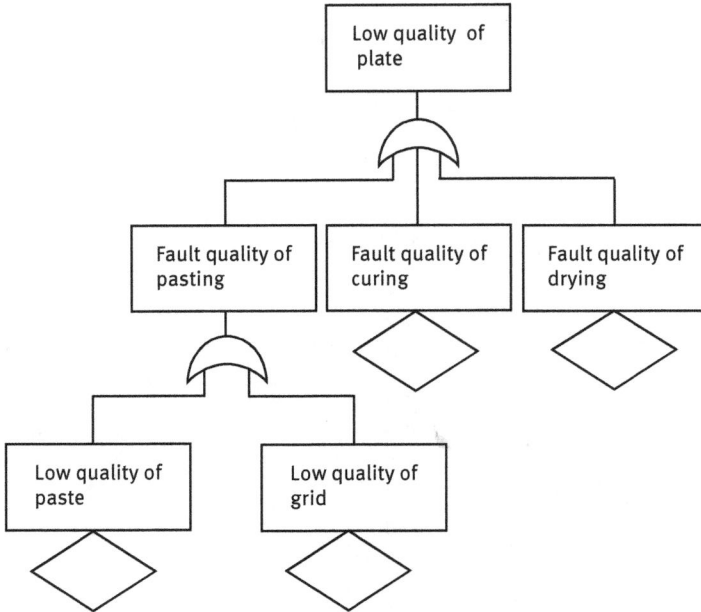

Fig. 8. Causal tree of low quality of lead acid battery plate in the manufacturing process.

3.2.1 Low quality of lead oxide

The low quality of lead oxide by the bill mill process is due to a rotational speed of ball mill beyond the range of 25–34 rpm, temperature of ball mill above 180° C or bad function of classifiers. The low quality of lead oxide according to the bill mill process is characterized by: amount of free lead outside the range 25–35 wt %, median pore diameter of particles beyond the range 1.67–2.67 μm, total pore volume of particles beyond the range 0.192–0.264 cm^3g^{-1} or specific surface area of particles outside the range 1.16–1.79 m^2g^{-1} [12, 14].

3.2.2 Low quality of paste

The low quality of the paste is caused by low quality of lead oxide, bad function of mixer, Acid to Oxide ratio A/O beyond the range 4.5–6 % or the wrong dosage of additives. The characteristics of the low quality of the paste are: crystal size above 2 μm, surface area of the crystal beyond the range 1.26 ∓0.06 m^2 g^{-1}, total volume pore of the crystal above 0.148 cm^3g^{-1}, moisture content of the paste outside the range of 11–12 %, density of the positive paste outside the range of 3.90–4.40 gcm^{-3}, density of the negative paste outside the range of 4.10–4.50 gcm^{-3} or amount of 1BS above 10 wt % [6, 7].

3.2.3 Low quality of grids

The possible causes of the low quality of the grid are: a wrong composition of the alloy, a bad function of the perforating machine or low geometry. The low quality of the grid is characterized by a very thick or very fine grid.

3.2.4 Fault quality of pasting

A fault quality of pasting is created by a bad function of pasting machines, low quality of grid, low quality of the paste or bad brushing. A bad function of pasting machines can cause by a faulty coating of paste, dross on the hopper or bad function of the hopper.

3.2.5 Fault quality of curing

The fault quality of curing is caused by a low recrystallization and interconnection of paste particles or a low grid oxidation and a low formation of corrosion layer under unsuitable conditions of humidity, temperature and time.

3.2.6 Fault quality of drying

The fault quality of drying is due to a bad control of temperature above 60°C, time below 12 h and relative humidity different to 100 %.

The fault quality of curing and drying is characterized by: low adhesion of the active material, specific surface area beyond the range 0.8-1 $m^2 g^{-1}$, size crystals above 10 μm, amount of free lead in the positive plates above 1 %, amount of free lead in the negative plates above 5 %, moisture content of the plate above 0.2 %, low cohesion or the presence of cracks [2, 6]. The detailed causal tree of the low quality of lead acid battery plate in the manufacturing process is shown in Fig. 9.

The basic causes that have constituted the origin of defects are:
–   Fault quality of pasting (low pressure press, bad function of hopper, dross on the hopper, bad brushing, low quality of paste or low quality of grid).
–   Fault quality of curing (low recrystallization and interconnection of paste particle, low grid oxidation and low formation of corrosion layer).
–   Fault quality of drying (relative humidity different to 100 %, drying time below 12 h, drying temperature above 60° C).

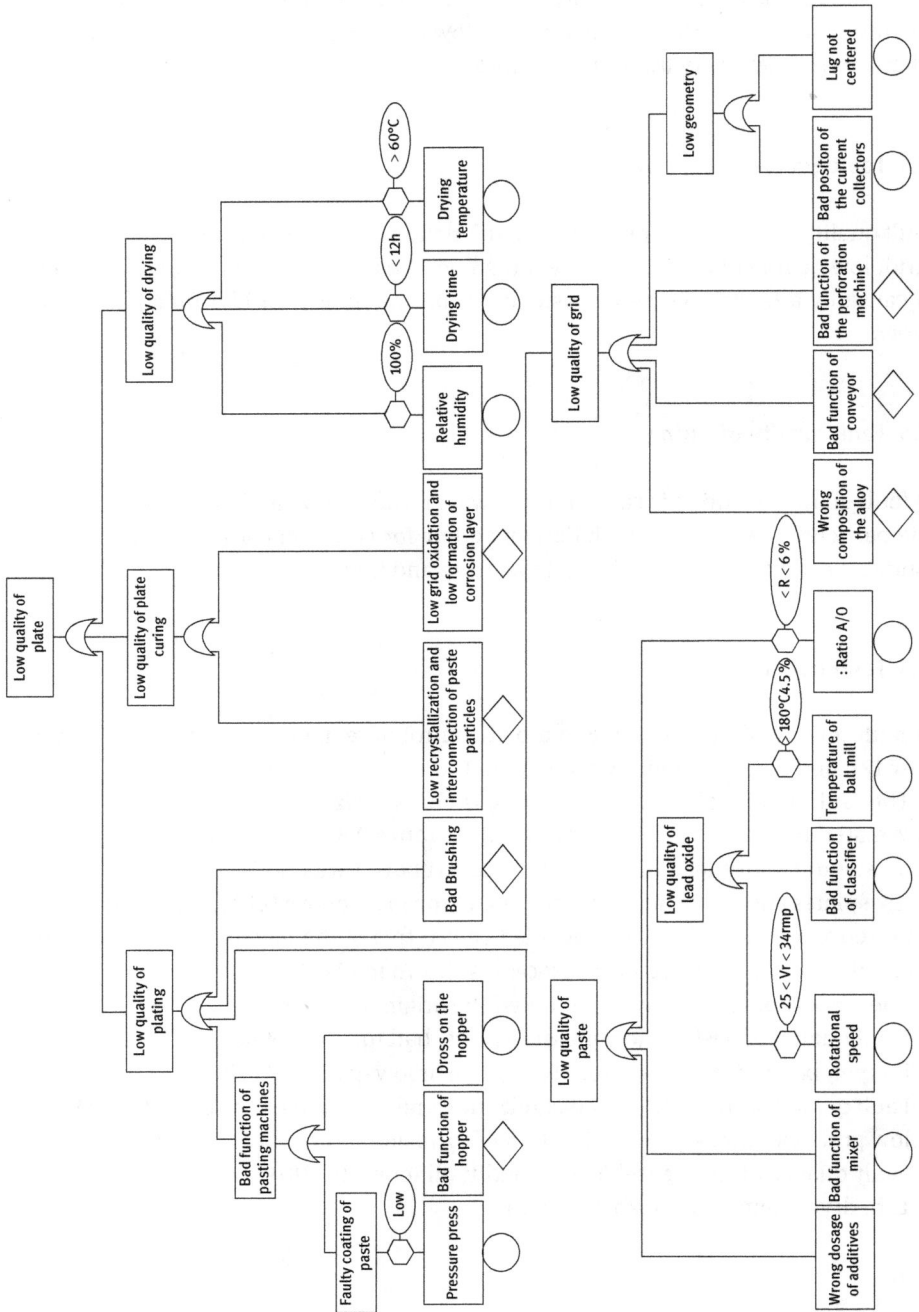

Fig. 9. Causal tree of low quality of lead acid battery plate in the manufacturing process.

4 Conclusion

In this paper, a study of the lead acid battery plate degradation during the manufacturing process is described using tools analysis of reliability such as SADT, Ishikawa diagram and CTA. This approach presents a SADT used to modeling the manufacturing processes of lead acid batteries plate (manufacturing of lead oxide, paste and grid, the process of pasting, curing and drying). The CTA is presented to find the origin of the plate degradation (low quality of lead oxide, paste, grid, or fault quality of pasting or curing and drying). Finally, this study presents a scientific enrichment for communion researchers; it allows adducing an improvement in the manufacturing process in order to increase the battery lifetime during operation process. As perspective of this study, the correlation between the degradation causes and parameters variation of the electrical equivalent model will be describe in order to determining the degradation depth of the lead acid battery during the manufacturing process.

Bibliography

[1] K. Brik and F. Ben Ammar. Causal tree analysis of depth degradation of the lead acid battery. *J. Power Sources*, 228:39–46, 2013.

[2] Detchko Pavlov. Lead Acid Batteries: Science and Technology. Elsevier Store, United Kingdom, 1st Edition, 1995.

[3] J.R. Pierson. Control of vital chemical processes in the preparation of lead-acid battery active materials. *J. Power Sources*, 158:868–873, 2006.

[4] D.A.J. Rand, D.P. Boden, C.S. Lakshmi, R.F. Nelson and R.D. Prengaman. Manufacturing and operational issues with lead-acid batteries. *J. Power Sources*, 107:280–300, 2002.

[5] Dahodwalla Huzefa and Sunil Heart. Cleaner production options for lead-acid battery manufacturing industry. *J. Cleaner Production*, 8:133–142, 2000.

[6] F. Storck. *Effect of compression and addition of additives on improving performance of a lead accumulator.*. PhD thesis, University Pierre and Marie CURIE , Paris, France, 2008.

[7] W.R. Kitchens, R.C. Osten and D.W.H. Lambert. Advances in manufacturing systems for the production of pastes for lead/acid battery plates. *J. Power Sources*, 53:263–267, 1995.

[8] D. McAllister, R. Ponraj, I. Francis Cheng and B. Edwards. Increase of positive active material utilization in lead-acid batteries using diatomaceous earth additives. *J. Power Sources*, 173:882–886, 2007.

[9] L.T. Lam, O. Lim, H. Ozgun, D.A.J Rand. Seeking enhanced lead/acid battery performance through the use of conductive tin-dioxide-coated glass-flakes. *J. Power Sources*, 48:83–111, 1994.

[10] K.R. Bullock. Lead/acid batteries. *J. Power Sources*, 51:1–17, 1994.

[11] N. Bagshaw. Lead alloys: past, present and future. *J. Power Sources*, 53:25–30, 1995.

[12] D.P. Boden. Improved oxides for production of lead-acid battery plates. *J. Power Sources*, 73:56–59, 1998.

[13] K. Brik and F. Ben Ammar. Reliability analysis of lead acid batteries based on fault tree and impedancespectroscopy. *European Journal of Electrical Engineering*, 13:91–129, 2010.

[14] L.T. Geyer. *The evaluation of potential improvements of barton pot oxides for lead acid batteries*. PhD thesis, Faculty of Applied Science, France, 2002.

Biographies

Raja Yahmadi was born in Tunis, Tunisia, on December 19, 1988. She obtained the research Master in Conversion and process of electrical energy in 2012 from the National High School of Engineer of Tunis ENSIT. Since 2013, she is currently with Research Laboratory Materials, Measurements and Applications, INSAT. Her research interests include electronics and power electronics.

Kais Brik was born in Tebourba, Tunisia on Mars 23, 1966. He received, the master degree and (DEA) from the Superior National School of Technical Education (ENSET) in 1991 and 1995 respectively. He received A PHD degree from The National Engineering School of Tunis-Tunisia (ENIT), in 2009. He has been a secondary school teacher from 1991 until 2000. From 2001 until 2009, he is teaching in the higher institute of technological studies of Rades, Tunisa. And from 2010 until now, he is teaching in the higher institute of multimedia arts of manouba (ISAMM), Tunisia. He is interested in the reliability analysis of electric power storage elements.

Faouzi Ben Ammar was born in Tunis, Tunisia, on May 15, 1962. He received the B. Eng. degree in electrical engineering from the National School of Engineers of Monastir (ENIM), Monastir, Tunisia, in 1987, and the D.E.A. and Ph. D. degrees from the National Polytechnic Institute of Toulouse (INPT), Toulouse, France, in 1989 and 1993, respectively. From 1993 to 1998, he was an Engineer with Alstom Company, Belfort, France, working on the development of control drive systems. In 1998, he joined the National Institute of Applied Sciences and Technology, Tunis, as an Assistant Professor, where since 2004, he has been Enabling at the Research Department and a Professor of power electronics. His research interests include power electronics and machine modeling, control induction motor drives, pulse width modulation, multilevel inverters and reliability.

A. Nayli, S. Guizani and F. Ben Ammar

Modeling and Analysis of the Open-End Stator Winding Permanent Magnet SM with Salient-Poles fed by VSI

Abstract: In this paper, the authors propose the mathematical model in Park reference frame of the open-end stator winding salient poles permanent magnet synchronous machine. The machine is supplied by two 2-level inverters and compared with classic permanent magnet synchronous machine. The feeding machine by two 2-levels cascaded inverters based on phase-disposition PWM strategy is presented. Comparative simulation analysis using the THD voltage, THD stator current, the torque ripple is shown.

Keywords: Open-End Stator Winding Permanent Magnet Synchronous Machine, Cascaded Inverters, THD Voltage, THD Current, Torque Ripple, Power Segmentation.

1 Introduction

In industrial applications, the association AC machines-inverters are most used such as railway traction, electric ship propulsion system, electrical vehicles systems, aerogenerator and aeronautics. The power segmentation is a good solution in order to increase power for that several research have been developed in inverter structures [1–5], or in AC machine structures including multiphase machines [6–9], multi star machines [7, 10, 11], and the open-end stator windings machines [12–17]. If the induction motors are designed for the rotation speeds of 750 to 3000 rpm, and therefore are not suited for the applications at reduced speed because their yield fall with the drop in speed, the synchronous machines have the advantage to operate with low and high speed. Moreover, these machines operate in a wide power range (from a few watts to several MW of power). In particular, the permanent magnet synchronous machine is increasingly used in variable speed applications for the benefits that it brings. Furthermore, this machine also has view multiple configurations at the level of these stator windings such as using multiple phases, or multiple stars [18, 19]. Currently, many research subjects are developed on the open-end stator windings having a

A. Nayli, S. Guizani and F. Ben Ammar: A. Nayli, University of Tunis, ENSIT Research Laboratory of Materials, Measurements and Applications MMA, INSAT, email: n.ayli@hotmail.fr, S. Guizani, University of El Manar, IPEIM, Research Laboratory of Materials, Measurements and Applications MMA, INSAT, email: guizanisami@yahoo.fr, F. Ben Ammar, National Institute of Applied Sciences and Technology, University of Carthage, email: faouzi.benamar@insat.rnu.tn

De Gruyter Oldenbourg, ASSD – Advances in Systems, Signals and Devices, Volume 7, 2018, pp. 105–122.
https://doi.org/10.1515/9783110470529-007

number of phases superior or equal to three [20, 21]. In the first part, the mathematical modeling of the open-end stator winding permanent magnet synchronous machine "OEWPMSM" with salient poles in the Park reference frame is described. The second part presents the supply of the proposed machine by two three-phase 2-level inverters based on PWM technique and compared with the classical synchronous machine using the simulation results of the THD voltage, THD stator current and torque ripple. Finally, the association of the machine with three-phase two-level cascaded inverters based on phase-disposition PWM strategy is presented for the improvement the THD voltage, THD stator current and torque quality.

2 Modeling the open-end stator winding permanent magnet synchronous machine with salient poles

The open-end stator winding permanent magnet synchronous machine "OEWPMSM" is usually studied in the (d, q) reference ($\omega(d,q) = \omega r$), the figure 1 represents the model of the machine.

The relation that links the flux and the currents is described by:

$$\begin{bmatrix} \Psi_{sd} \\ \Psi_{sq} \end{bmatrix} = \begin{bmatrix} L_{sd} & 0 \\ 0 & L_{sq} \end{bmatrix} \begin{bmatrix} i_{sd} \\ i_{sq} \end{bmatrix} + \begin{bmatrix} \Psi_f \\ 0 \end{bmatrix} \tag{1}$$

with: Ψ_f: The flux of the permanent magnets by pole.

The voltage of the machine is related by the following matrix:

$$\begin{bmatrix} V_{sd1} - V_{sd2} \\ V_{sq1} - V_{sq2} \end{bmatrix} = \begin{bmatrix} R_s & 0 \\ 0 & R_s \end{bmatrix} \begin{bmatrix} i_{sd} \\ i_{sq} \end{bmatrix} + \frac{d}{dt} \begin{bmatrix} \Psi_{sd} \\ \Psi_{sq} \end{bmatrix} + \begin{bmatrix} 0 & -\omega_{dq} \\ \omega_{dq} & 0 \end{bmatrix} \begin{bmatrix} \Psi_{sd} \\ \Psi_{sq} \end{bmatrix} \tag{2}$$

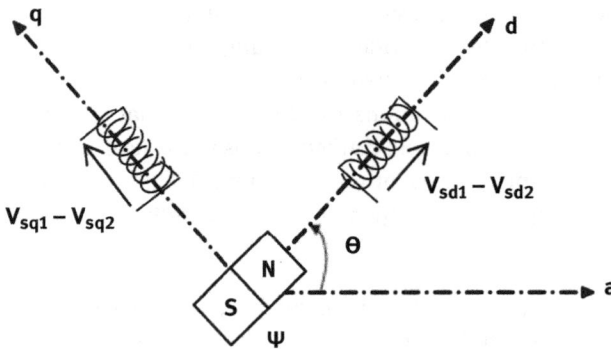

Fig. 1. Representation of the machine in the (d, q) reference frame.

where R_s is the resistance of the stator, L_d, L_q are the inductance of the stator, and d- and q- axis respectively.

If the "OEWPMSM" is supplied by two voltage sources, the mathematical current model is written in (d,q) reference frame, and described by the following state equation representation:

$$\frac{d}{dt}[I] = [A].[I] + [B].[V] \tag{3}$$

with:

$X(t) = \begin{bmatrix} i_{sd} & i_{sq} \end{bmatrix}^T$: the state vector

$U(t) = \begin{bmatrix} V_{sd1} - V_{sd2}, & V_{sq1} - V_{sq2}, & \Psi_f \end{bmatrix}^T$: the control vector

Matrices $[A]$ and $[B]$ are:

$$[A] = \begin{bmatrix} \dfrac{-R_s}{Lq} & \omega_{dq}\dfrac{Lq}{Ld} \\ -\dfrac{Ld}{Lq} & \dfrac{-R_s}{Lq} \end{bmatrix} ; \quad [B] = \begin{bmatrix} \dfrac{1}{Ld} & 0 & 0 \\ 0 & -\dfrac{1}{Lq} & \omega_{dq}\dfrac{1}{Lq} \end{bmatrix}$$

The following expression represents the electromagnetic torque of the salient poles permanent magnet synchronous motor:

$$T_{em} = \frac{3}{p}[(L_d - L_q)i_{sd}i_{sq} + \Psi_f i_{sq}] \tag{4}$$

3 Supply of the open-end winding permanent magnet synchronous machine

The simulation model is validated for a supply based on PWM strategy by 2-level and 3-level inverters for a power machine P = 40 kW.

3.1 Supply by three phase 2-level inverter

The "OEWPMSM" is fed by three-phase 2-level inverter to each entry of the stator winding. This structure is represented by figure 2.

To supply the "OEWPMSM" by two voltage source inverters, we used a carrier PWM technique. The figure 3 shows the principle of carrier PWM to supply the open-end winding synchronous machine by four 2-level inverters.

Figure 4 shows the reference voltage Vref$_{11}$ and Vref$_{21}$ with the triangular signal for the control of the two two-level inverters. Figure 5 shows the voltage (V$_{s11}$-V$_{s12}$), (V$_{s21}$-V$_{s22}$) and phase-to-phase machine voltage (U$_1$) which is 3 levels to supply with two three-level inverters, with:

- V$_{s11}$, V$_{s12}$ simple voltage of inverter A_1
- V$_{s11}$ −V$_{s12}$ pole voltage of inverter A_1

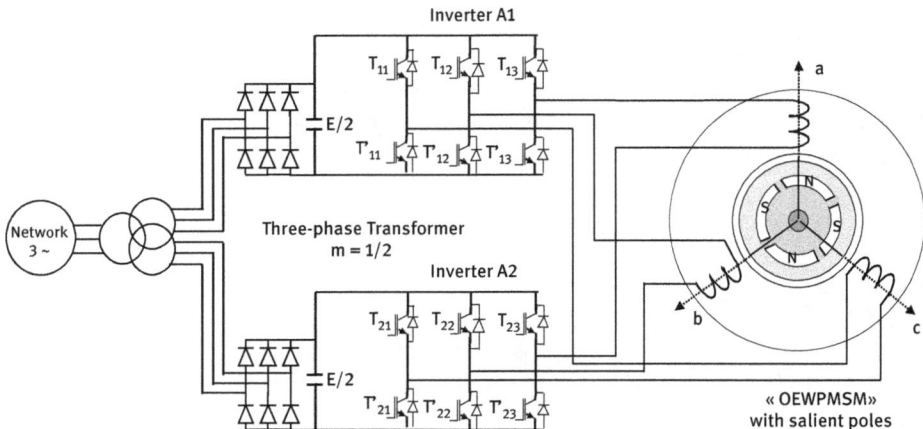

Fig. 2. Supply the "OEWPMSM" by two 2-level inverters.

- V_{s21}, V_{s22} simple voltage of inverter A_2
- $V_{s21} - V_{s22}$ pole voltage of inverter A_2
- $U_1 = (V_{s11} - V_{s12}) - (V_{s21} - V_{s22})$ pole voltage of the machine.

To view the performance of the machine "OEWPMSM", we will also simulate the permanent magnet synchronous machine with star winding "PMSM", and then compare the performance of the two machine structures.

The evolution of the stator currents, speed and the torque for the "PMSM" and "OEWPMSM" are shown by respectively figures 6 and 7.

In order to analyze the torque ripples, we defined ΔT_{em} by the expression:

$$\Delta T_{em} = 100 \times \frac{T_{Max} - T_{moy}}{T_{moy}} \tag{5}$$

Figure 8 shows the enlarging effect of the torque of the "PMSM" during the permanent mode for a load torque $T_r = T_n$.

Then we can calculate the torque ripple of the "PMSM" for this operation mode.
$$\Delta T_{em} = \frac{259 - 180}{180} 100 = 43.88\%$$
Figure 9 shows the enlarging effect of the torque of the "OEWPMSM" during the permanent. Then we can calculate the torque ripple of the "OEWPMSM" for this operation mode. $\Delta T_{em} = \frac{217.4 - 180}{180} 100 = 20.77\%$

The "OEWPMSM" clearly improves the torque ripple compared with "PMSM".

Figure 10 shows the waveform and the harmonic content of the phase-to-phase machine voltage for the "PMSM" with THD voltage = 70.01%.

Figure 11 shows the waveform of the phase-to-phase machine voltage and the harmonic content of the voltage for "OEWPMSM" with THD voltage = 44.15%. It is clear that this machine "OEWPMSM" offers a better level of voltage between two phases, improves the THD voltage and extends the band-width.

$$Vref_{11} = V_{Max}/2 \sin(2\pi f_s t)$$

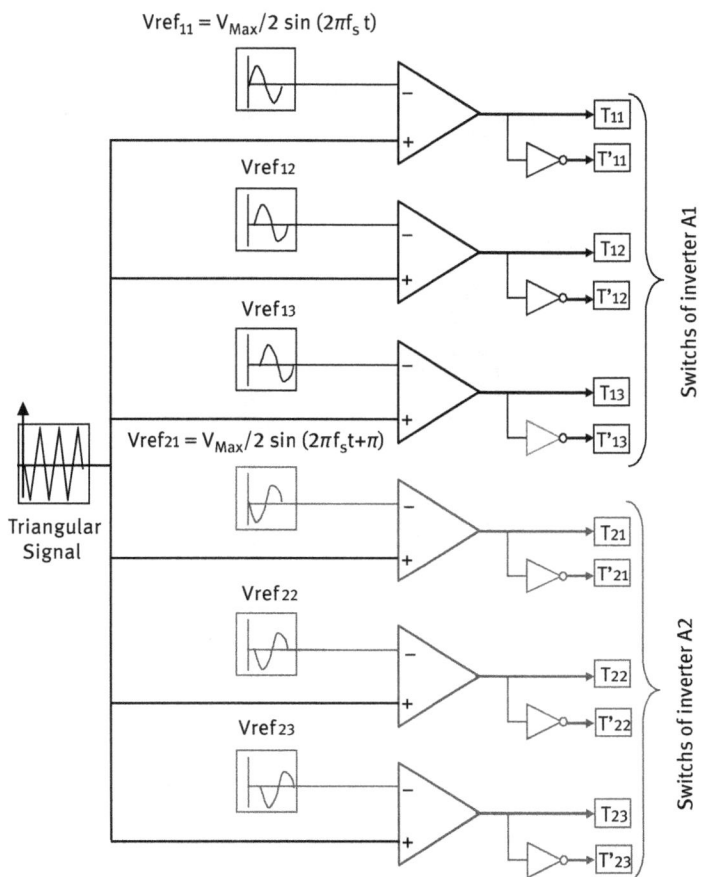

Fig. 3. Principle of the PWM sine triangle.

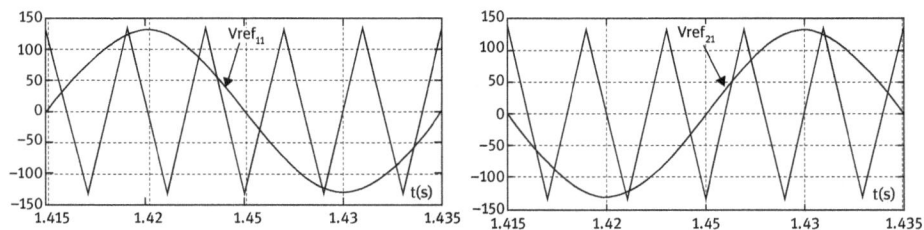

Fig. 4. Signal of $Vref_{11}$ and $Vref_{21}$ with a triangular signal for the control of inverter A1 and A2.

Figure 12 shows the evolution of the stator current of "PMSM" during the permanent mode. Figure 13 shows the waveform and the harmonic content of the stator current for the machine "PMSM". With THD current = 17.80 %.

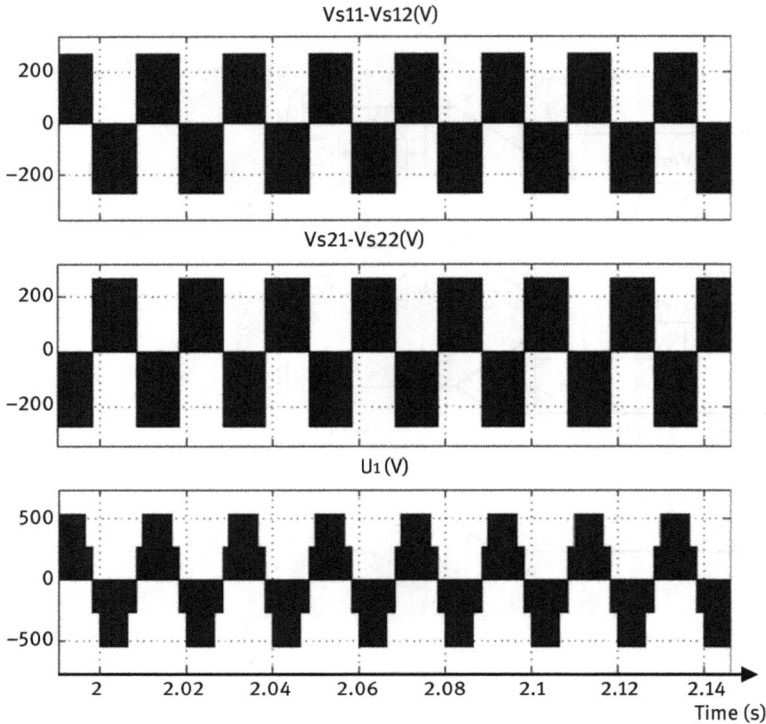

Fig. 5. Pole voltage inverter and phase-to-phase machine voltage.

The figure 14 shows the evolution of the stator current during the permanent mode. Figure 15 shows the waveform and the harmonic content of the stator current for the machine "OEWPMSM", With THD current = 5.76 %.

We note the important advantage which presents the the machine "OEWPMSM" compared to the machine "PMSM" to the THD stator current.

3.2 Supply by two 2-level cascaded inverters

The "OEWPMSM" is supplied by two converters A1 and A2, as shown by figure 16. The converter is formed by cascading two 2-level inverters.

We used the phase-disposition PWM strategy to control the two three phase 2-level cascaded inverters, the three reference signals of frequency f_m = 50 Hz and amplitude Am are compared with two triangular carriers of frequency f_p = 5000 Hz and amplitude $\frac{A_p}{p}$, as shown by figure 17.

Figure 18 shows two triangular carriers and a signal reference for controlled inverters 11 and 12.

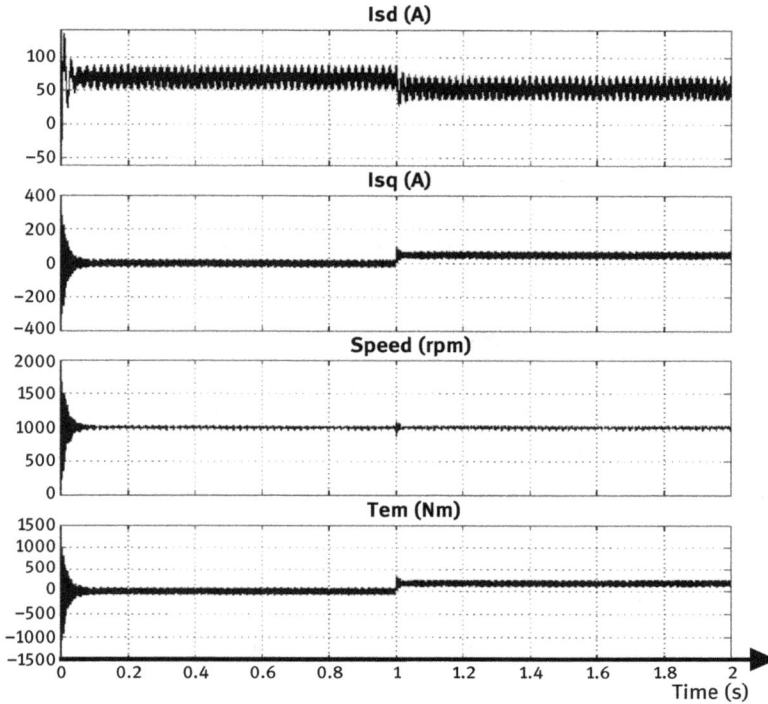

Fig. 6. Evolution of the current, speed and torque for "PMSM".

Figure 19 shows the voltage (V_{s11}-V_{s12}), (V_{s21}-V_{s22}) of converters A_1 and A_2 and phase-to-phase machine voltage U_1 which is 5 levels to supply with two three-level inverters. Figure 20 shows the evolution of the stator currents, speed and the torque.

Figure 21 shows the enlarging effect of the torque of the "OEWPMSM" during the permanent mode.

Then:

$$\Delta T_{em} = \frac{199.2 - 180}{180} \times 100 = 10.67\,\%$$

Figure 22 shows the waveform of the phase-to-phase machine voltage and the harmonic content of the voltage with THD voltage = 26.91 %.

Figure 23 shows the evolution of the stator current during the permanent mode. The waveform and the harmonic content of the stator current for the machine is shown by figure 24, with THD current = 4.01 %.

The association salient poles "OEWPMSM" with 2-level cascaded inverters improves the torque quality, offers better THD voltage and THD stator current and increases the levels of voltage. The characteristics of the machine for nominal power $P = 40$ KW.

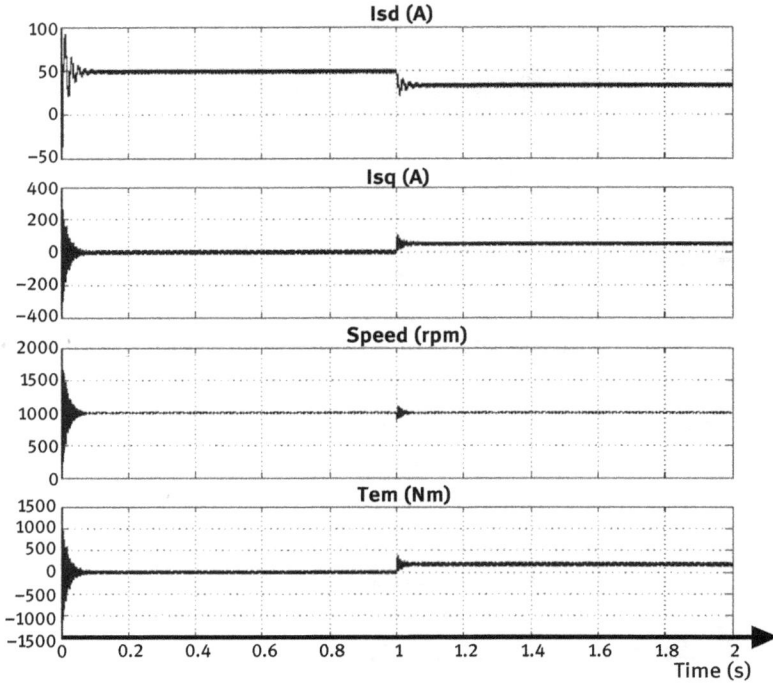

Fig. 7. Evolution of the current, speed and torque for "OEWPMSM".

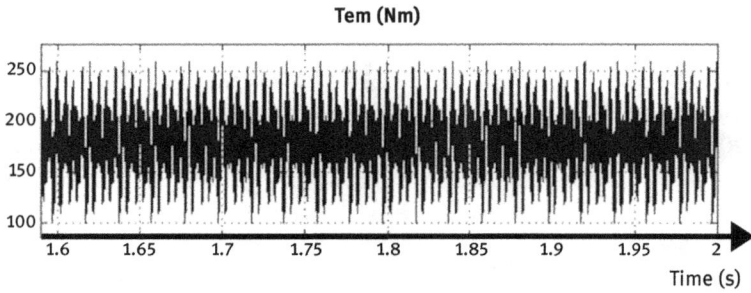

Fig. 8. Enlarging effect of the waveform torque "PMSM".

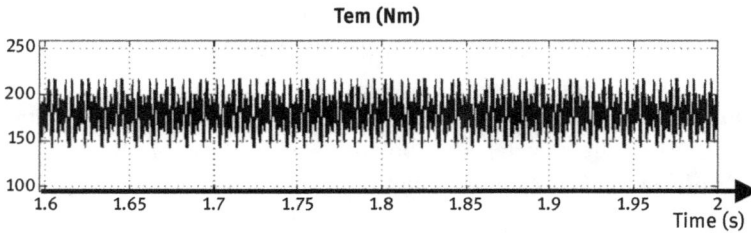

Fig. 9. Enlarging effect of the waveform torque "OEWPMSM".

Signal to analyze
○ Display selected signal   ○ Display FFT window

U1 (V)

FFT analysis

Fundamental (50Hz) = 460.3 , THD = 70.01%

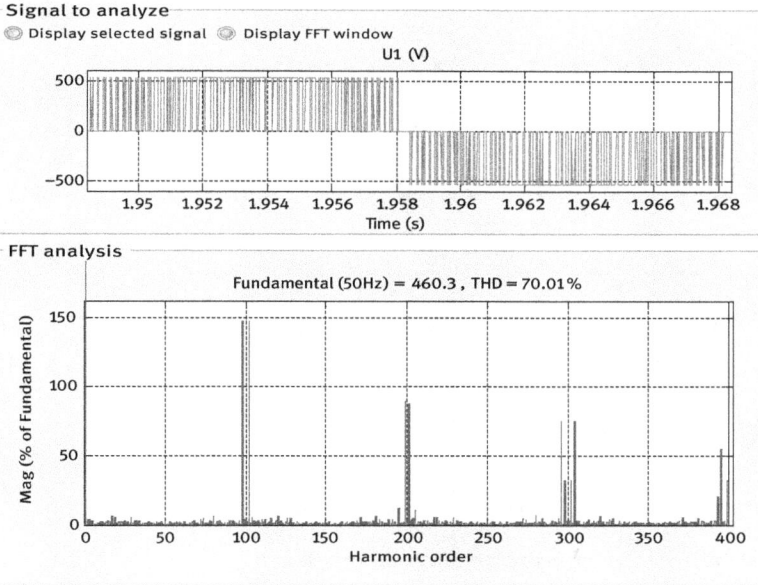

Fig. 10. Waveform and harmonic ration of machine voltage for "PMSM".

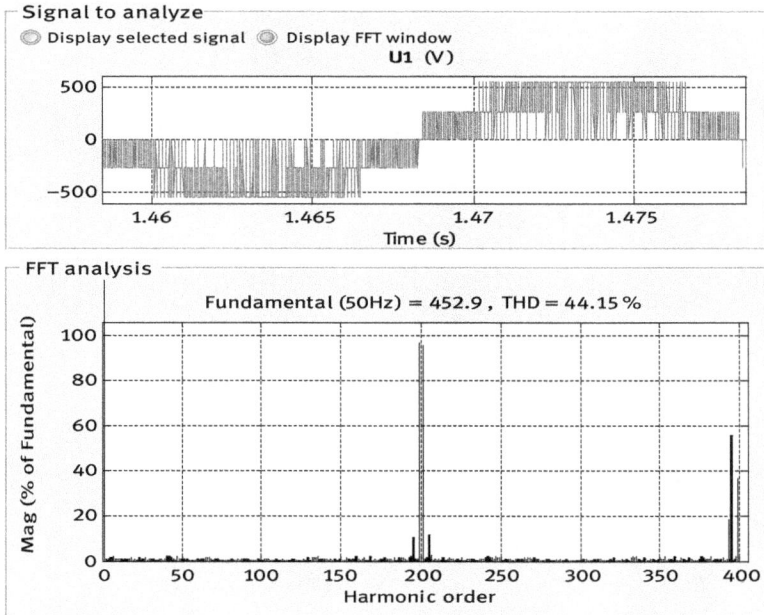

Signal to analyze
○ Display selected signal   ○ Display FFT window
U1 (V)

FFT analysis

Fundamental (50Hz) = 452.9 , THD = 44.15 %

Fig. 11. Waveform and harmonic ration of machine voltage for "OEWPMSM".

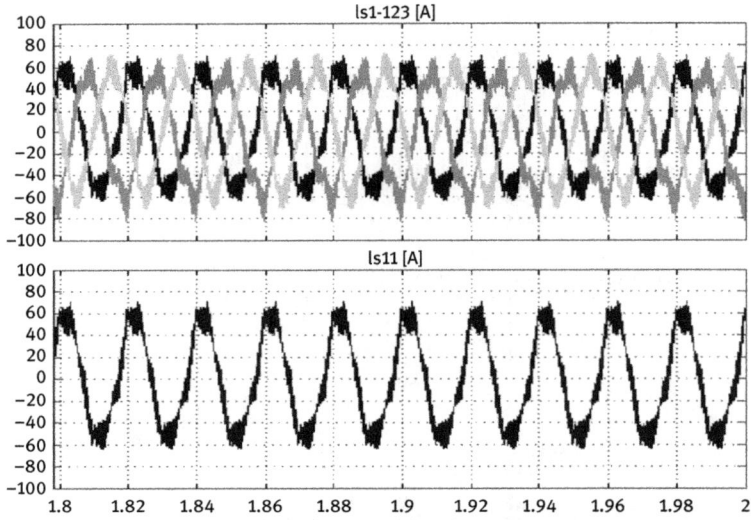

Fig. 12. Evolution of the current during the permanent mode for "PMSM".

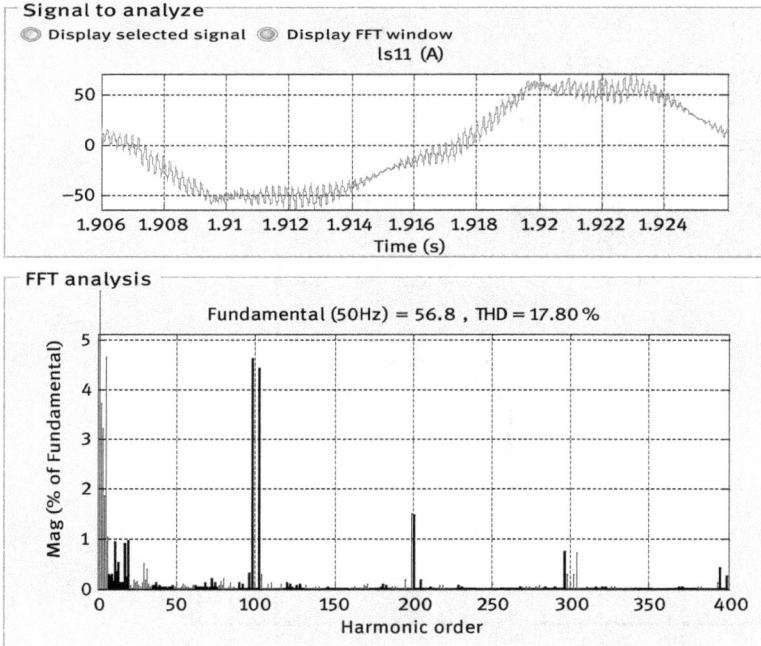

Fig. 13. Waveform and harmonic ration of current for "PMSM".

Is1-123 [A]

Is11 [A]

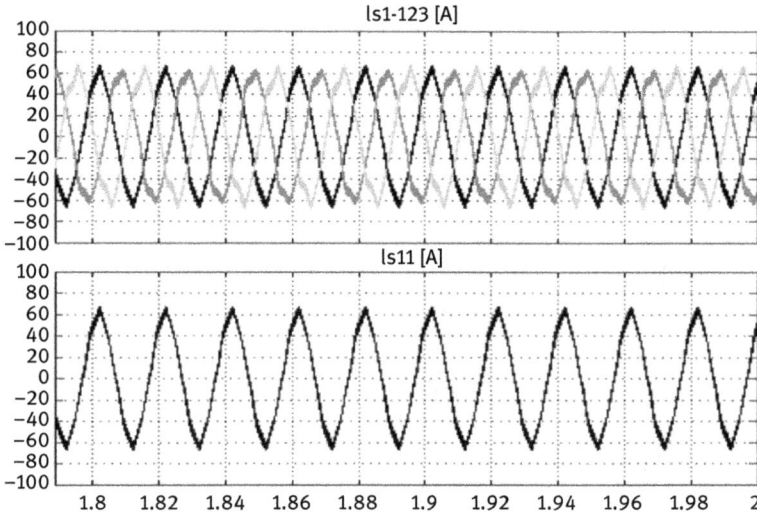

Fig. 14. Evolution of the current during the permanent mode for "OEWPMSM".

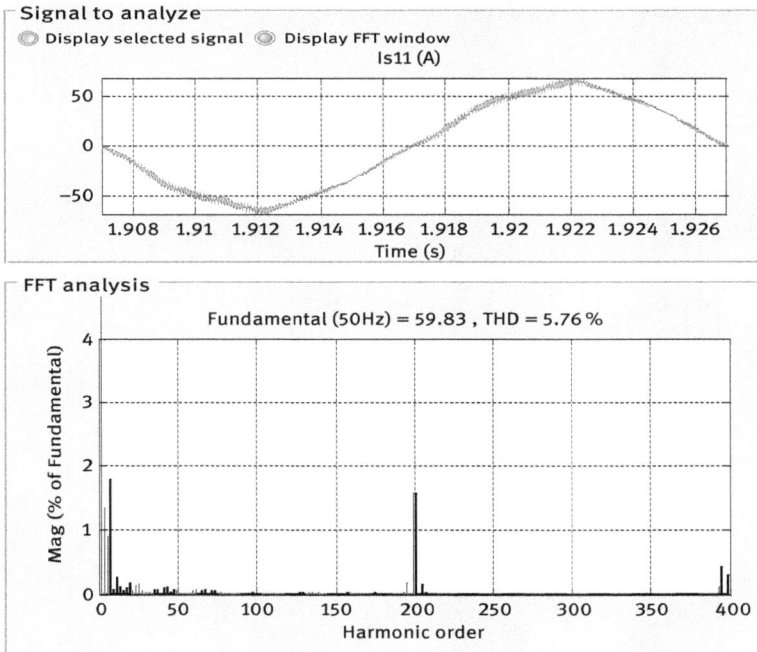

Signal to analyze
◉ Display selected signal ◉ Display FFT window
Is11 (A)

Time (s)

FFT analysis

Fundamental (50Hz) = 59.83 , THD = 5.76 %

Mag (% of Fundamental)

Harmonic order

Fig. 15. Waveform and harmonic ration of current for "OEWPMSM".

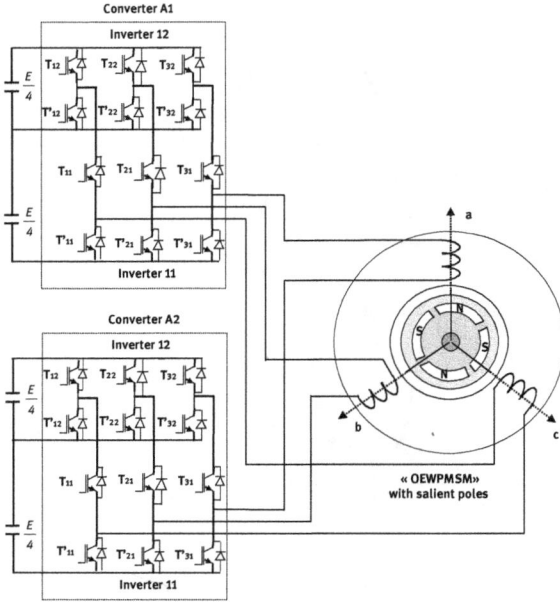

Fig. 16. Supply the "OEWPMSM" by two 2-level cascaded inverters.

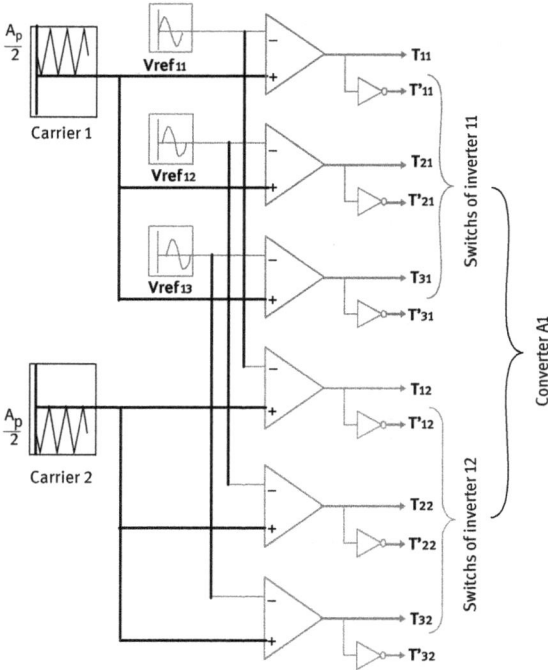

Fig. 17. Principle of the phase-disposition PWM for control two three-phase cascaded inverters.

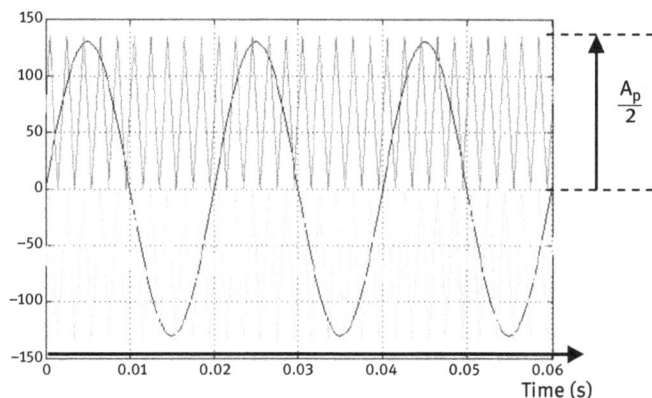

Fig. 18. Signal reference and 2 carriers vertically shifted.

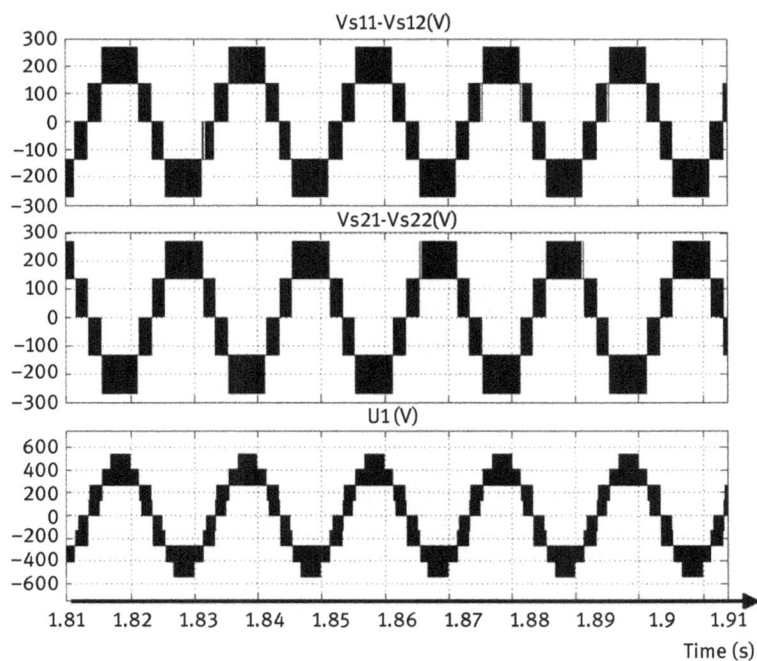

Fig. 19. Pole voltage inverter and phase-to-phase machine Voltage.

Speed $n = 1000\,\text{tr/min}$
Resistance of stator $R_s = 0.065\,\Omega$.
Inductance of stator $L_d = 0.655\,\text{mH}$

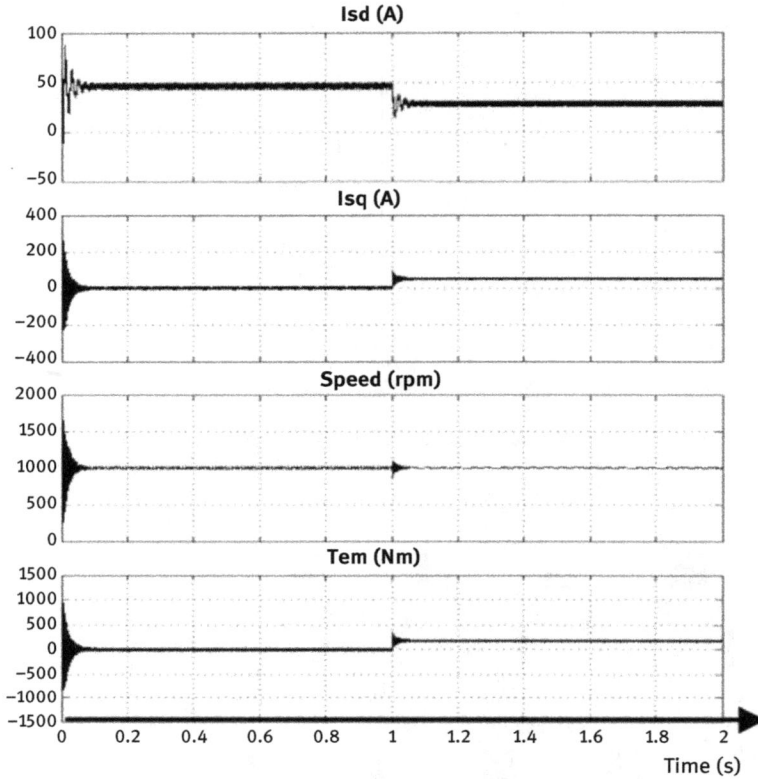

Fig. 20. Evolution of the current, speed and torque "OEWPMSM".

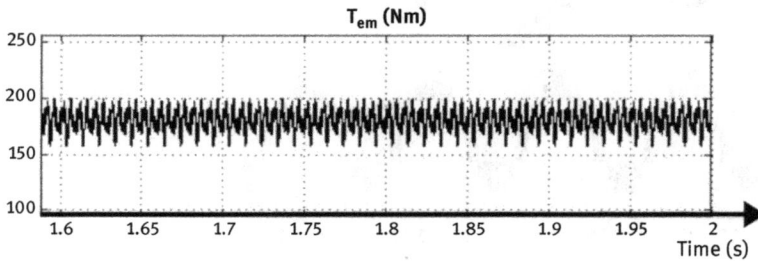

Fig. 21. Enlarging effect of the waveform torque "OEWPMSM".

Inductance of stator L_q= 0.575 mH
Magnet flux Ψ_f = 0.14 Wb
Inertia moment J = 0.01 kg m^2
Viscous force f = 2×10^{-3} Nms/rad

Signal to analyze

○ Display selected signal ○ Display FFT window

U1 (V)

FFT analysis

Fundamental (50Hz) = 450.9 , THD = 26.91%

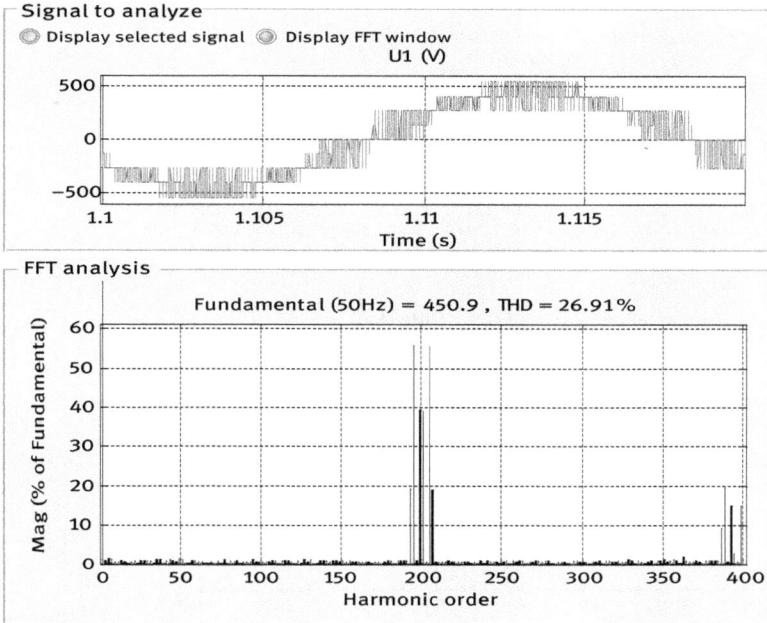

Fig. 22. Waveform and harmonic ration of machine voltage "OEWPMSM".

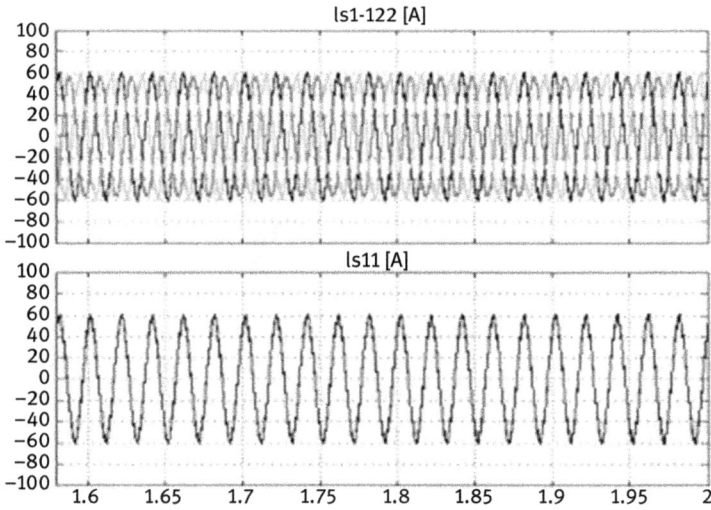

ls1-122 [A]

ls11 [A]

Fig. 23. Evolution the current during the permanent mode.

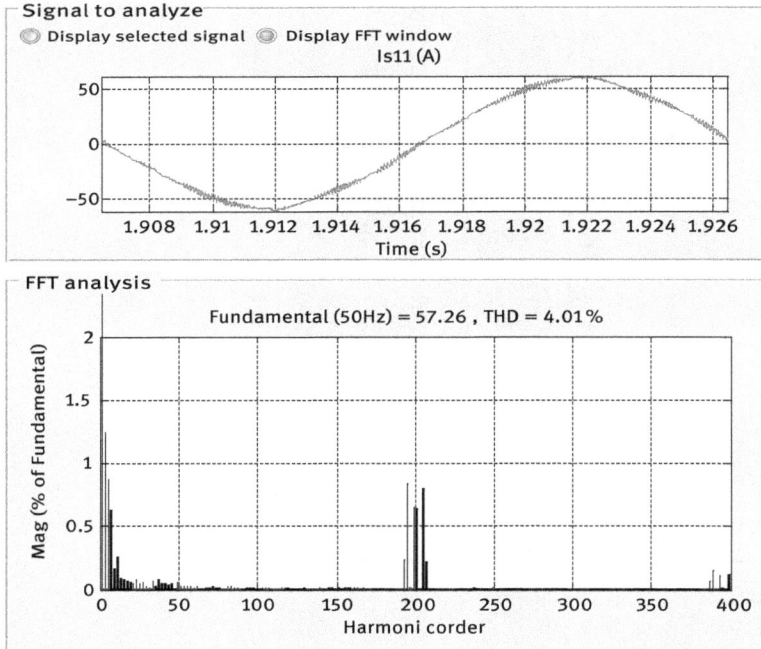

Fig. 24. Waveform and harmonic ration of current "OEWPMSM".

4 Conclusion

The open-end stator winding permanent magnet synchronous machine with salient poles is modeled in the reference (d, q). The supply of the proposed machine by three-phase 2-level inverters and three phase 2-level cascaded inverters is presented.

A THD analysis of the voltage and current and the ripple rate of the torque are presented with comparison between the "OEWPMSM"and classic "PMSM". The feeding of the "OEWPMSM" by 2-level inverters offers many advantages compared with classic "PMSM". Indeed, the better THD voltage and THD current, extends the band-width, quality of torque and reduced dimensions of the inverters of the half power, and the possibility of the operation in degraded mode.

The association of the "OEWPMSM" with the 2-level cascaded inverters allows to improve the performances. Indeed, that shows a better THD voltage, best quality of the torque and especially more degrees of liberty in degraded mode.

Bibliography

[1]   L.M. Tolbert, Z.P. Fang and T.G. Habetter. Multilevel converters for large electric drives. *IEEE Trans. on Industry Applications*, 35, 1999.

[2]   J. Rodriguez, J.-S. Lai and F.Z. Peng. Multilevel inverters: A survey of topologies, controls, and applications. *IEEE Trans. on Industrial Electronics*, 49(4):724–738, 2002.

[3]   J. Rodriguez, S. Bernet, B. Wu, J.O. Pontt and S. Kouro. Multilevel voltage-source-converter topologies for industrial medium-voltage drives. *IEEE Trans. on Industrial Electronics*, 54(6):2930–2945, 2007.

[4]   L.G. Franquelo, J. Rodriguez, J.I. Leon, S.Kouro, R. Portillo and M.A.M. Prats. The Age of Multilevel Converters Arrives. *IEEE Industrial Electronics Magazine*, 49(4):28–39, 2008.

[5]   S. Kouro, M. Malinowski, K. Gopakumar, J. Pou, L.G. Franquelo, BinWu, J. Rodriguez, M.A. PÃ©rez and J.I. Leon. Recent Advances and Industrial Applications of Multilevel Converters. *IEEE Trans. on Industrial Electronics*, 57:2553–2580, 2010.

[6]   G.K. Singh, V. Pant and Y.P. Singh. Voltage source inverter driven multi-phase induction machine. *Journal of computer and Electrical Engineering Elsevier*, 29:813–834, 2003.

[7]   S.N. Vukosavic, M. Jones, E. Levi and J. Varga. Rotor flux control of symmetrical six phase induction machine. *Electric Power Systems Research, Elsevier*, 75:142–152, 2005.

[8]   M. Jones, S.N. Vukosavic, D. Dujic and E. Levi. A Synchronous Current Control Scheme for Multiphase Induction Motor Drives. *IEEE Trans. on Energy Conversion*, 24(4), 2009.

[9]   A.S. Abdel-Khalik, M.I. Masoud and B.W. Williams. Vector controlled multiphase induction machine: Harmonic injection using optimized constant gains. *Electric Power Systems Research, Elsevier*, 89:116–128, 2012.

[10] S. Guizani and F. Ben Ammar. The eigenvalues analysis of the double star induction machine supplied by redundant voltage source inverter. *Int. Review of Electrical Engineering IREE*, 3:300–311, 2008.

[11] K. Marouani, L. Baghli, D. Hadiouche, A. Kheloui and A. Rezzoug. A New PWM Strategy Based on a 24-Sector Vector Space Decomposition for a Six-Phase VSI-Fed Dual Stator Induction Motor. *IEEE Trans. on Industrial Electronics*, 55(5):1910–1920, 2008.

[12] V.T. Somasekhar, M.R. Baiju and K. Gopakumar. Dual two-level inverter scheme for an open-end winding induction motor drive with a single DC power supply and improved DC bus utilization. *IEEE Proc.-Electr. Power Appl*, 151, 2004.

[13] G. Mondal, K. Gopakumar, P.N. Tekwani and E. Levi. A Reduced-Switch- Count Five-Level Inverter with Common-Mode Voltage Elimination for an Open-End Winding Induction Motor Drive. *IEEE Trans. on Industrial Electronics*, 54:2344–2351, 2007.

[14] S. Srinivas and V.T. Somasekhar. Space-vector-based PWM switching strategies for a three-level dual-inverter-fed open-end winding induction motor drive and their comparative evaluation. *IET Electric Power Applications*, 2:19–31, 2008.

[15] A. Nayli, S. Guizani and F. Ben Ammar. Open-end winding induction machine supplied by two flying capacitor multilevel inverters. *ICEESA, Hammamet, Tunisia*, 2013.

[16] S. Guizani, A. Nayli and F. Ben Ammar. Fault-Tolerant control for open-end stator winding induction machine supplied by two three phase cascaded inverters with one failed inverter. *Journal of Electrical Engineering Jee*, 14, 2014.

[17] A. Nayli, S. Guizani and F. Ben Ammar. Implantation of scalar control for the open-end winding induction machine on FPGA Spartan 3E. *Journal of electrical system JES*, 11, 2015.

[18] S. Kallio, M. Andriollo, A. Tortella and J. Karttunen. Decoupled d-q Model of Double-Star Interior-Permanent-Magnet Synchronous Machines. *IEEE Trans. on industrial electronics*, 60(6), 2013.

[19] H. Yashan, Z. Zi-Qiang and L. Kan. Current control for dual three-phase permanent magnet synchronous motors accounting for current unbalance and harmonics. *IEEE Journal of emerging and selected topics in power electronics*, 2, 2014.

[20] P.S. Alexandru, M. Fabien, K. Xavier, S. Eric and A. Bruyere. Flux-weakening operation of open-end winding drive integrating a cost-effective high-power charger. *IET Electrical system in transportation*, 3:10–21, 2013.

[21] Y. Lee and J. Ha. Hybrid Modulation of Dual Inverter for Open-End Permanent Magnet Synchronous Motor. *IEEE Trans. on Power Electronics*, 30(6), 2015.

Biographies

Abdelmonoem Nayli was born in Metlaoui, Tunisia on September 06, 1985. He received, in 2010, the master ' s degree in the higher school of sciences and techniques of Tunis (ESSTT) and a (DEA) in 2012. He is interested in the power electronics and machine modelling, control induction Motor drives and Reliability.

Sami Guizani was born in Gaafour, Tunisia on January 27, 1966. He received, in 1990, the master⁄s degree in the higher national school of technical studies (ENSET) and certificate of a higher specialized studies (CESS) in 2000. A (DEA) from The National Engineering School of Tunis-Tunisia (ENIT), in 2003.He has been a secondary school teacher from 1990 until 1998. And from 2003, he is teaching in the higher institute of technological studies of Rades, Tunisa. He joined the IPEIEM, University of El Manar in 2009 as assistant professor. He is interested in the power electronics and machine modelling, control induction Motor drives and Reliability. Actually, he is preparing a doctoral thesis dealing with the double star asynchronous machine.

Faouzi Ben Ammar was born in Tunis, Tunisia on May 15, 1962. He received the Engineer degree in Electrical engineering from National Engineering School of Monastir-Tunisia (ENIM), in 1987, and (DEA) and the PHD degree from the National polytechnic Institute of Toulouse, France (INPT, ENSEEIHT) in 1989 and 1993 respectively. Dr. Ben Ammar occupied from 1993 to 1998 engineer ⁄ s post in Alstom company in France – Belfort in the development of control drives systems. He joined the National Institute of Applied sciences and Technology of Tunis (INSAT) in 1998 as assistant professor. Since 2004 he has been HDR and professor of power electronics at the same Institute. He is interested in the power electronics and machine modeling, control induction Motor drives, Pulse Width Modulation, Multilevel inverter, active filter, Reliability and RMAS analysis.

R. Koubaa and L. Krichen

A FC/UC Hybrid Source Energy Management Algorithm with Optimal Parameters

Abstract: This paper investigates the energy management problem of a hybrid power source with a Fuel Cell (FC) as a main energy source and an Ultra-Capacitor (UC) as a power buffer for an electric vehicle. The power split as well as the charge sustaining of the UC bank, which is an optimal control problem, are fulfilled by Ant Colony Optimization (ACO) algorithm. A preliminary study is performed in order to find the optimal parameters of the ACO algorithm. Two standard driving cycles (ECE 15) and (EUDS) are adopted for the simulation work while respecting the energetic and dynamic constraints of the hybrid source.

Keywords: Fuel Cell, Ultra-Capacitor, Energy Management, Charge Sustaining, Ant Colony Optimization, Optimal Parameters.

1 Introduction

Novel vehicular technologies such plug-in hybrid electric vehicles and hybrid electric vehicles are gaining an increasing interest as green alternatives instead of classical internal combustion engine vehicles. This work focuses in particular on Fuel Cell (FC) hybrid electric vehicles with an Ultra-Capacitor (UC) bank as an energy buffer. Hydrogen FCs represent the best non pollutant energy source for vehicular applications, in particular, Proton Exchange Membrane Fuel Cells, owing to its high power density and its lower temperature range compared with the other technologies [1, 2]. Some works adopt a battery stack as a secondary energy source, while others adopt a combination of both battery stack and UCs [3, 4]. The choice of UCs in this work is based on its dynamic characteristics that complement the performance of a fuel cell. Mostly, an UC is characterized by its fast dynamics that can respond for both instantaneous power demand and regenerative braking power. An UC has also a long life cycle and a high efficiency considering that it stores electrical energy without any conversion process [5, 6]. All these features enhance the FC performance that is characterized by its low dynamics which is incompatible with the fast variations of the demanded power and improve the overall energy efficiency. For a Fuel Cell / Battery based hybrid electric vehicle, Pourhashemi and al. applied Ant Colony Optimization to optimal energy management, and Yan Li and al. used optimal fuzzy power management [7, 8]. While

R. Koubaa and L. Krichen: University of Sfax, Tunisia, Emails: kb.rayhanel@gmail.com
lotfi.krichen@enis.rnu.tn

De Gruyter Oldenbourg, ASSD – Advances in Systems, Signals and Devices, Volume 7, 2018, pp. 123–138.
https://doi.org/10.1515/9783110470529-008

for the energy management of a Fuel Cell/ Ultra-capacitor hybrid power system, Song Lin and al. [9] used an adaptive-optimal control method, while authors in [10] adopted a rule based method. Dynamic programming is also a robust global optimization technique that was adopted in [11] for the energy management of a FC/UC/battery hybrid source. The power system architecture of the Fuel Cell/Ultra-Capacitor hybrid electric vehicle is presented in Fig. 1.

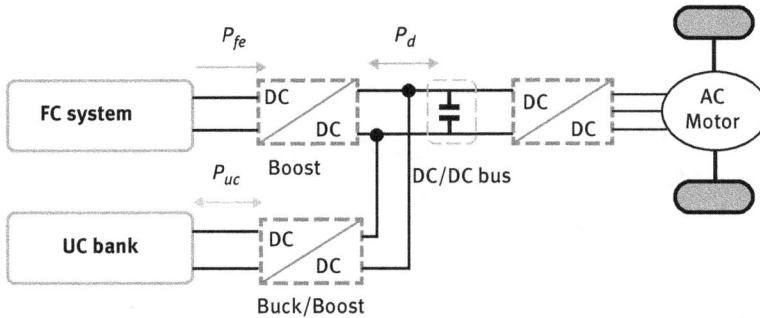

Fig. 1. Hybrid vehicle architecture.

In this paper, Ant Colony System (ACS), which is an algorithm derived from the Ant Colony Optimization (ACO) metaheuristic, is adopted to solve the energy management of the hybrid power source. The charge sustaining of the UC bank state of charge which represents an optimal control problem with final state constraints is performed as well with respect of the entire dynamic and energetic constraints imposed on the system. The ACO metaheuristic is introduced in the first section and the optimal values of the algorithm parameters are determined. The formulation of the optimization problem is presented in the second section, and the simulation results are presented in the final section.

2 Ant colony optimization algorithm

2.1 Introduction to the ACO metaheuristic

Ant Colony Optimization is originally introduced by the Italian scientist Marco DORIGO in 1991 within his PhD thesis, who is also one of the founders of the Swarm Intelligence. Ant Algorithms are inspired by the real behavior of ants, more preciously by its behavior while searching food. While walking from food source to the nest, each ant deposit a chemical substance called pheromone which can be smelled by other agents forming a pheromone trail, when choosing their way, ants tend to follow by probability the path containing more pheromone deposit than others as shown in Fig. 2 [12]. Eventually, after several cycles, the shortest path will converge [13, 14]. The main idea is the indirect communication among colony agents, called artificial ants, via pheromone trails.

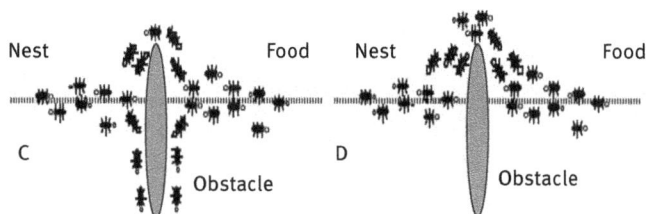

Fig. 2. Ant choosing the shortest path.

The first developed AC algorithm called Ant System (AS) has been applied to the TSP to test its effectiveness. Subsequently, many improvements have been proposed for this basic algorithm to reach a stronger exploitation of the solution space in order to guide the research process of the agents. Some improved versions are Max-Min Ant System (MMAS) algorithm, Ant-Q algorithm, Rank-based Ant System algorithm, and Ant Colony System (ACS) algorithm which is used in this work. In order to achieve the energy sharing between the FC and the UC bank, ACS algorithm is proposed. The AC algorithms developed recently are based on population approach and have been successfully applied to several N-P hard combinatory optimization problems, such as the benchmark problem Traveling Salesman Problem (TSP) [15]. So this work brings out the applicability of the ACS algorithm in solving dynamic optimization problems, more preciously, Optimal Control Problems (OCP) [16]. In fact, the optimal control theory aims to lead the system from an initial given state to a certain final state, while minimizing a cost function and meeting certain criteria [17]. In this paper, the State Of Charge (SOC) of the UC bank is chosen to be the state variable in the OCP.

The indirect communication of ants is performed by artificial pheromone trail which is a kind of distributed numeric information that will be modified and updated after each single step in order to reflect the required new experience until the convergence of the algorithm. The state transition rule adopted by the ACS algorithm is a probability distribution expressed by the following equation which gives a probability of an ant to move from a vertex to another:

$$P_{ij}^k = \frac{\tau_{ij}^\alpha}{\sum_{l \in T} \tau_{ij}^\alpha} \tag{1}$$

P_{ij}^k is the probability of ant k in action i to choose action j
α is a weighting factor
τ_{ij} is the pheromone trail between action i and action j
T is the set of possible actions of ant k

The initial pheromone deposit is initialized with a small positive value τ_0. After each step, the ants construct their solutions by adding pheromone on every chosen trail until the completion of a full cycle; this operation is called the local pheromone update:

$$\tau_{ij}^k(k+1) = (1-\rho)\tau_{ij}^k(k) + \rho\tau_0 \tag{2}$$

ρ is the evaporation parameter which represents the speed of pheromone evaporation of the previous trials and it can be adjusted for limiting or expanding the solutions space. Subsequently after making a complete cycle (iteration), the pheromone trail is updated globally depending on the quality of the solution. This global update happens only for the best solution vertexes and it represents an improvement and a characteristic of ACS algorithm compared with other ant colony algorithms:

$$\tau_{best_{ij}}^k(k+1) = (1-\rho)\tau_{best_{ij}}^k(k) + \rho\sum J_{best}^{-1} \tag{3}$$

J is the cost function of ant *k*.

2.2 Impact of the ACO parameters variation

In this section, a preliminary study is performed in order to adopt the optimal values of the ACO parameters later in the simulation works. The performance of the algorithm including the precision and the speed of convergence will be observed while changing the values of the parameters First, the optimal number of ants m will be extracted. The number of ants is quietly the most important parameter since they represent the agents that perform the pheromone update as indicated before. So the value of m should be selected appropriately to guarantee the convergence of the algorithm to the desired value, which is in this study, the initial value of the UC state of charge fulfilling the charge sustaining strategy. According to Fig. 3, the optimal value of ants is m = 100. For values of m that are inferior to 100, the algorithm does not converge to the optimal solution and the error is a small value having a maximum percentage equal to 1.04

Another important parameter is the evaporation coefficient ρ which influences strongly the convergence of the algorithm.The evaporation coefficient is a small

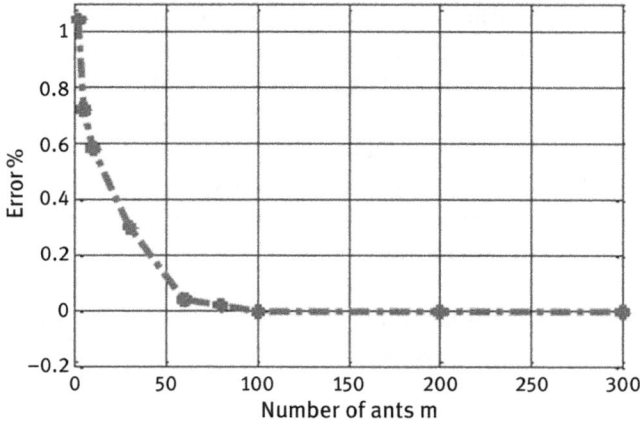

Fig. 3. Impact of the number of ants m on the convergence of the algorithm.

positive value between 0 and 1; it translates the speed of the pheromone evaporation in each vertex at instant k, so it influences the pheromone update at the next instant $k + 1$. This parameter is adopted to avoid the rapid convergence of the algorithm to a suboptimal solution.

Figure 4 demonstrates the impact of the value of the evaporation coefficient on the performance of the algorithm for fixed values of m and the number of cycles K_{max}. According to the figure, the optimal solution with 0 % error is achieved for the value of

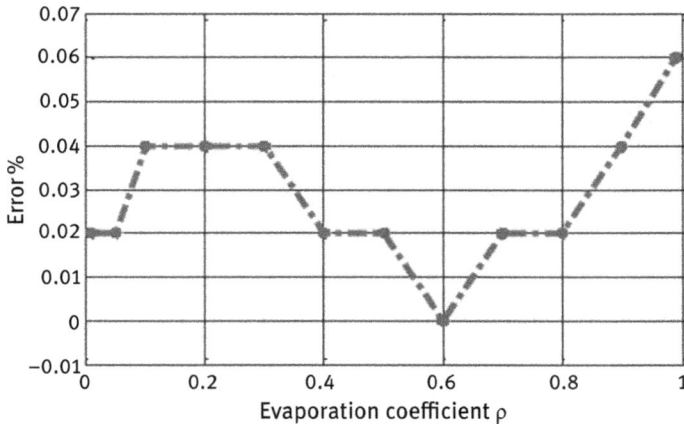

Fig. 4. Impact of the evaporation coefficient on the convergence of the algorithm.

ρ equal to 0.6. For too low values of ρ that are inferior to 0.01, the error is equal to 0.02 % which is considered as a minor value. It means that low values of the evaporation coefficient induce the stagnation of old solutions and prevent the algorithm from exploring new solution spaces; hence, it converges to a local optimum. For high values of ρ that are superior to 0.8, the value of the error reaches 0.06 %. It means that a strong evaporation of the pheromone deposit cancels the experience acquired by ants from previously performed cycles which attenuates the inter-communication of ants and ruins the whole concept of the ACO metaheuristic.

For a fixed value of the number of ants m, the variation of the evaporation coefficient influences also the speed of the algorithm convergence as shown in Fig. 5. The best performance is achieved within 15 cycles with a coefficient equal to 0.6. For lower values of ρ, the number of cycles ranges between 15 and 25. Whereas for the values of ρ that are superior to 0.6, the number of cycles reaches 50, which is relatively high compared with the number of cycles achieved under low values of the coefficient. So considering the speed of convergence criterion, a low evaporation is better than a strong one since this latter induces a loss of the information acquired during the previous steps, but in return, it may lead to a suboptimal convergence.

3 Energy management of the hybrid source

3.1 Hybrid source modeling

The demanded power which will be shared between the FC system and the UC bank is calculated with the vehicle parameters shown in Table 1. For the formulation of the

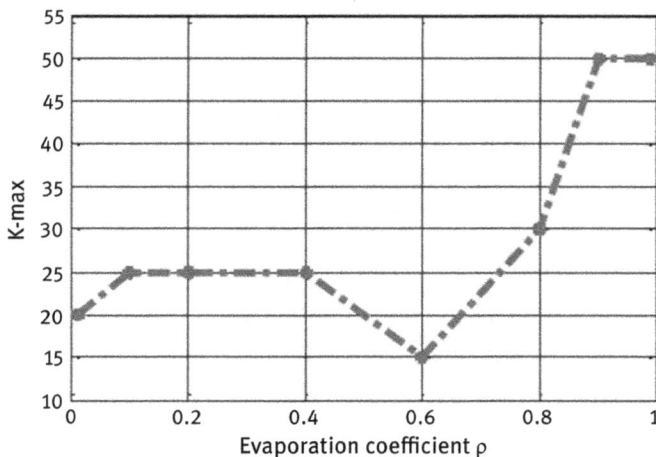

Fig. 5. Impact of the evaporation coefficient on the speed of convergence.

Tab. 1. Vehicle parameters.

Weight	850 kg
Frontal surface	2.59 m²
Drag coefficient	0.37
Coefficient of rolling resistance	0.0136

energy management problem, the FC system power P_{fc} is the control input. For the ACO algorithm, it is simply required to present a matrix containing the possible values of the FC power, bounded by the limit values. The step size is selected depending on the desired precision and the margin between the two extreme limits of the power value, while the slow dynamics of the FC system are implemented implicitly. The gradient of the FC power ΔP_{fc} is put into consideration by restricting the choice of the next value $P_{fc}(k+1)$ compared with the current value $P_{fc}(k)$, ΔP_{fcfall} for the falling rate limit and ΔP_{fcrise} for the rising rate limit.

Unlike the FC system, the UC model must be implemented explicitly on the algorithm, as an electric model. The adopted model of the UC in this work is a basic model as shown in Fig. 6 [15]. The SOC of the UCs is a state variable of the system; hence, all the dynamic equations including the state equation have to be clearly developed in the algorithm.

$$SOC(t) = \frac{Q(t)}{Q_{max}} \tag{4}$$

Q(t) is the instantaneous charge of the UC bank Q_{max} is the maximal charge of the UC bank

According to the Fig. 6, P_{uc} is the power going through the capacity, while P_{ucnet} is the net power passing through the terminal nodes of the UC cell. The power is

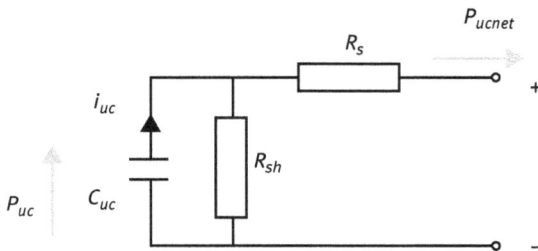

Fig. 6. Electric model of an UC.

counted negative during the discharge and positive during the charge mode. The serial resistance R_s, representing a non negligible loss is put into consideration, while R_{sh} represents a negligible parallel loss. P_{uc} is then expressed in terms of the state variable SOC and the serial resistance R_s as following:

$$P_{uc} = \left(1 - \sqrt{1 - \frac{2R_s - P_{ucnet}}{E_{max} \times SOC}}\right)\frac{E_{max} \times SOC}{R_s \times C_{uc}} \tag{5}$$

P_{ucnet} is the net power stored or provided by the UC bank
P_d is the demanded power
R_s is the total serial resistance of the UC bank
C_{uc} is the total capacitance of the UC bank
E_{max} is the maximal energy of the UC bank

3.2 Problem formulation

The state variable of the energy management problem is the SOC of the UC:

$$x(k) = SOC(k) = \frac{Q_{uc}(k)}{Q_{max}} \tag{6}$$

The control input is the FC system power P_{fc} and the cost function to be minimized is:

$$J = M|SOC(N) - SOC(1)| \tag{7}$$

M is a weighting parameter.

The optimal control in this algorithm consists in leading the state of the system from a known initial value $SOC(1)$ to a certain finial value $SOC(N)$. In this paper, we have adopted the charge sustaining strategy of the UC bank which means $SOC(1) = SOC(N)$, so that in every new cycle the sufficiency of the State Of Charge is well guaranteed. The state equation of the system is described by the following equation:

$$SOC(k+1) = SOC(k) - P_{uc}(k)\frac{\Delta T}{E_{max}} \tag{8}$$

where: k is the discrete index, N is the number of samples, and ΔT is the discretization step.

According to the performance of the ACS, the algorithm will generate the value of the control input $P_{fc}(k)$ following a probability function that depends on amount of the pheromone deposit described previously by (1). In each step, the choice of the action or the state variable is limited by certain values owing to the constraints of the system, mainly, the slow dynamics of the FC system and the limit boundaries of the UC bank state of charge. This dynamic change of the solutions space highlights the applicability of ACO approach for solving dynamic optimization problems, which is normally applied to static optimization. At each moment, the power produced by the FC system and the UC bank must be equal to the power demanded by the vehicle, whether it is positive (demand) or negative (braking). According to the modeling and sizing work, the chosen fuel cell must have a maximum power $P_{fcmax} = 20kW$, while the minimum power value is $P_{fcmin} = 0.2kW$. We have also considered in this work the slow dynamic of the FC system, hence, the power gradient of the FC must be limited by a negative value ΔP_{fcfall} corresponding to the falling rate of the power and a positive value ΔP_{fcrise} corresponding to the rising rate of the power. The high value of the UC bank State Of Charge must be limited by the upper value $SOC_{max} = 0.9$ for security reasons, while the low values must not be under the lower limit $SOC_{min} = 0.2$ to avoid the severe diminution of the buck-boost converter efficiency when the voltage of the UC bank is too low. The equality constraint insures the equality of the demanded and provided power at each instant to prevent any losses:

$$g = P_{fc}(k) + P_{ucnet}(k) - P_d(k) = 0 \tag{9}$$

The inequality constraints are:

$$h_1 = P_{fc}(k) - P_{fcmin} \geq 0 \tag{10}$$

$$h_2 = P_{fcmax} - P_{fc}(k) \geq 0 \tag{11}$$

$$h_3 = P_{fc}(k) - P_{fc}(k-1) - \Delta P_{fcfall}\Delta T \geq 0 \tag{12}$$

$$h_4 = \Delta P_{fcrise}\Delta T - P_{fc}(k) - P_{fc}(k-1) \geq 0 \tag{13}$$

$$h_5 = SOC(k) - SOC_{min} \geq 0 \tag{14}$$

$$h_5 = SOC_{max} - SOC(k) \geq 0 \tag{15}$$

4 Simulation results

The ACS algorithm is implemented and tested under different conditions. In this work, two standard driving cycles were used, the first one is the $(ECE15)$ that shows city driving conditions and characterized mostly by a low speed, and the second is the $(EUDC)$ driving cycle which represent the extra urban driving conditions. The velocity profiles are presented in Fig. 7 and Fig. 8.

The preliminary study performed earlier in this paper, generated the optimal values of the number of ants m, the evaporation parameter and the maximum number of cycles K_{max} that are adopted in the simulation to enhance the performance of the algorithm. The power distribution of both driving cycles is shown in Fig. 9 and Fig. 10, respectively. When the power demand is constant, the fuel cell power P_{fc} is generated exactly equal to the load, so the UC bank power P_{uc} is null. Also, the variation of the fuel cell power P_{fc} respects well the lower and upper variation rate limits $[-0.30.3]$ kW for the $(ECE15)$ cycle and $[-30.8]$ kW for the $(EUDC)$ cycle. This constant state of

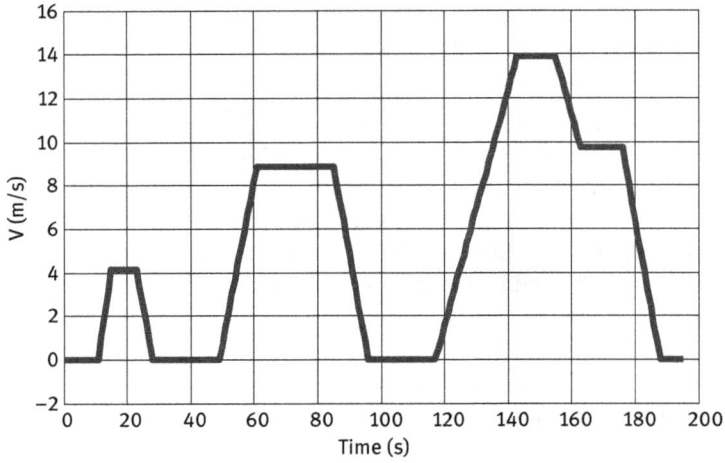

Fig. 7. Velocity profile (ECE15).

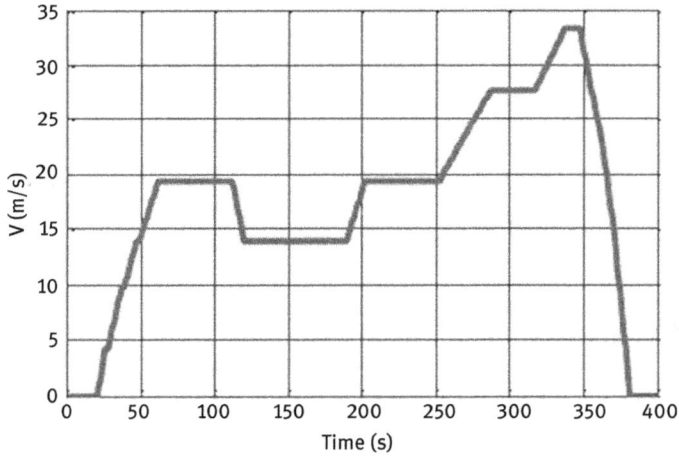

Fig. 8. Velocity profile (EUDC).

the fuel cell power enhances its performance and enables a smooth dynamic of the system, which obviously will extend its lifetime.

As noticed, the UC bank interferes mostly when the power demand varies instantly, thanks to its high power density, or when the load variation rate is too high compared with the slow response of the fuel cell. UCs enabled also the absorption of the entire regenerative energy of braking which enhances the energetic performance of the hybrid source.

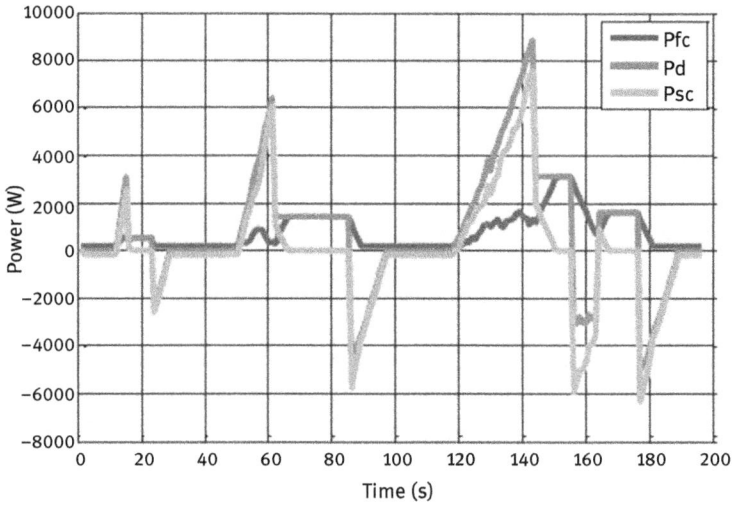

Fig. 9. Power distribution (ECE15).

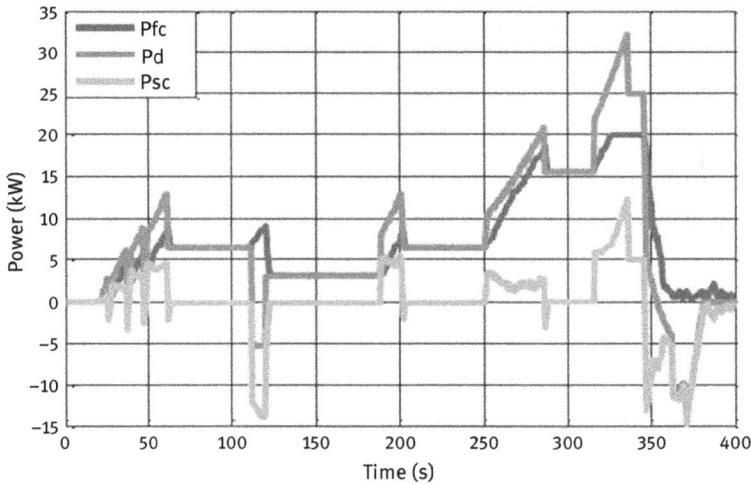

Fig. 10. Power distribution (EUDC).

Figure 11 and Fig. 12 show the instantaneous equality of the demanded power P_d and the provided power $P_{fc} + P_{uc}$ fulfilling the equality constraint for both cycles, which contributes in minimizing the energetic losses and improves the overall efficiency. The UC bank state of charge of each cycle is presented in Fig. 13 and Fig. 14, respectively. The upper and lower limits [0.20.9] are clearly well respected, and the algorithm

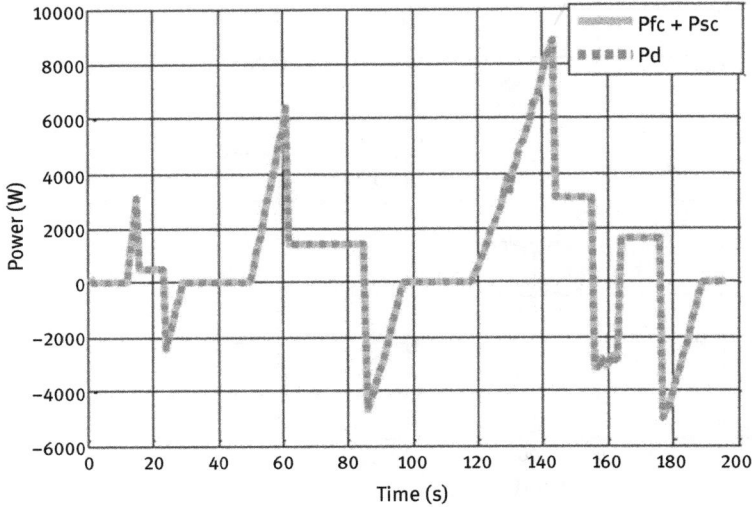

Fig. 11. Demanded and provided powers (ECE15).

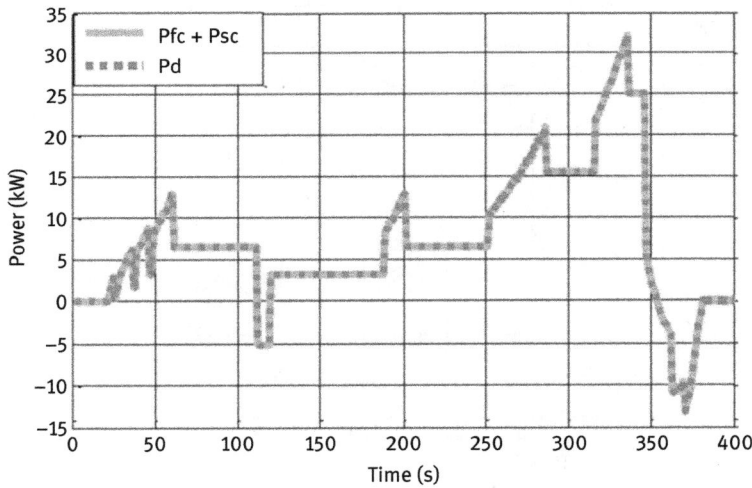

Fig. 12. Demanded and provided powers (EUDC).

converges well to the desired final value of the state of charge fulfilling the charge sustaining. The error of the final state of charge when using non optimal parameters ranges between 0.1 and 1.5%, while with optimal values we obtain an exact equality of the initial and final sate of charge.

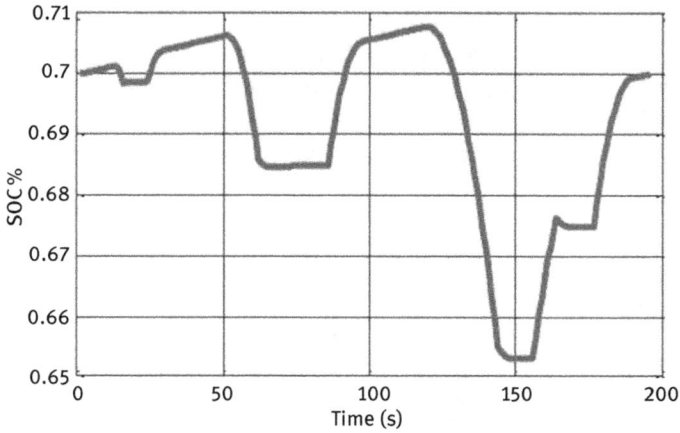

Fig. 13. SOC variation (ECE15).

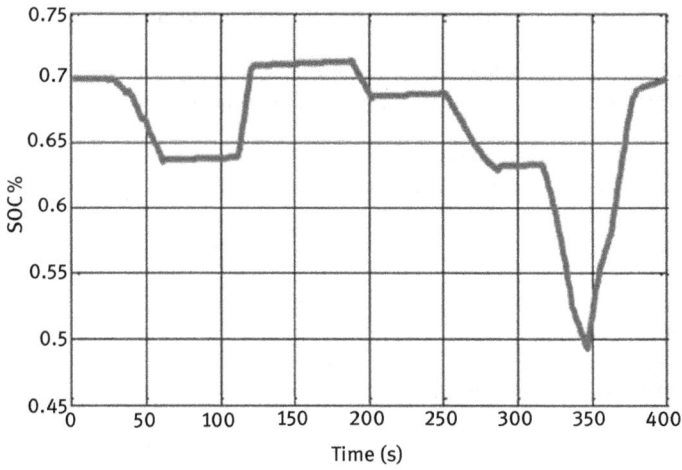

Fig. 14. SOC variation (EUDC).

5 Conclusion

The performance of the ACO metaheuristic in solving the energy management of a FC/UC power source, which is dynamic optimization problem, has been highlighted in this work. Furthermore, the charge sustaining of the UC bank state of charge has been successfully fulfilled. In order to ensure a convergence with a 0% error in the final state of charge value, a preliminary study has been held to enable the extraction of the optimal values of the ACS algorithm parameters. Both driving cycles (ECE15) and ($EUDC$) validated the performance of the algorithm with respect of the entire equality and inequality constraints that consider mostly the dynamic nature of each energy source.

Bibliography

[1] J. Wee. Applications of proton exchange membrane fuel cell systems. *Renewable and Sustainable Energy Reviews*, 11:1720–1738, 2007.

[2] O. Erdinc, B. Vural, M. Uzunoglu and Y. Ates. Modeling and analysis of an FC/UC hybrid vehicular power system using a wavelet-fuzzy logic based load sharing and control algorithm. *Int. Journal of Hydrogen Energy*, 34:5223–5233, 2009.

[3] X. Li , L. Hua, X. Lin, J. Li and M. Ouyang. Power management strategy for vehicular-applied hybrid fuel cell/battery power system. *Journal of Power Sources*, 191:542–549, 2009.

[4] M. Uzunoglu and M.S. Alam. Dynamic modeling, design and simulation of a PEM fuel cell/ultra-capacitor hybrid system for vehicular applications. *Energy Conversion and Management*, 48:1544–1553, 2007.

[5] Z. Yua, D. Zingera and A. Bose. An innovative optimal power allocation strategy for fuel cell, battery and supercapacitor hybrid electric vehicle. *Journal of Power Sources*, 196:2351–2359, 2011.

[6] S.F. Tie and Ch.W. Tan. A review of energy sources and energy management system in electric vehicles. *Renewable and Sustainable Energy Reviews*, 20:82–102, 2013.

[7] A. P. Pourhashemi and M. Ansarey. Ant Colony Optimization Applied to Optimal Energy Management of Fuel Cell Hybrid Electric Vehicle. In *IV Int. Congress on Ultra-Modern Telecommunications and Control Systems*, 2012.

[8] Ch. Y. Lia and G.P. Liu. Optimal fuzzy power control and management of fuel cell/battery hybrid vehicles. *Journal of Power Sources*, 192:525–533, 2009.

[9] W.S. Lin and Ch.H. Zheng. Energy management of a fuel cell/ultracapacitor hybrid power system using an adaptive optimal-control method. *Journal of Power Sources*, 196:3280–3289, 2011.

[10] M. Uzunoglu and M.S. Alam. Dynamic modeling, design and simulation of a PEM fuel cell/ultra-capacitor hybrid system for vehicular applications. *Energy Conversion and Management*, 48:1544–1553, 2007.

[11] M. Ansarey, M. Panahi, H. Ziarati and M. Mahjoob. Optimal energy management in a dual-storage fuel-cell hybrid vehicle using multi-dimensional dynamic programming. *Journal of Power Sources*, 250:359–371, 2014.

[12] M. Dorigo and L.M. Gambardella. Ant colonies for the traveling salesman problem. *TR/IRIDIA/*, 1996–3.

[13] M. Dorigo, G. Di Caro and L.M. Gambardella. Ant Algorithms for Discrete Optimization. *Artificial Life*, 5:137–172, 1999.

[14] M. Dorigo, M. Birattari and Th. Stutzle. Ant Colony Optimization: Artificial Ants as a Computational Intelligence Technique. *IRIDIA*, 2006.

[15] Th. Stutzle and M. Dorigo. ACO algorithms for the Travelling Salesman Problem. *IRIDIA*, Year !!!.

[16] J. van Ast, R. Babuska and B. De Schutter. Ant colony optimization for optimal control. *IEEE Congress on Evolutionary Computation*, Hong Kong, June 2008.
[17] E. Todorov. Optimal Control Theory. *Bayesian Brain, Doya, K. (ed), MIT Press*, 2006.
[18] P. Sharma and T.S. Bhatti. A review on electrochemical double-layer capacitors. *Energy Conversion and Management*, 51:2901–2912, 2010.

Biographies

Rayhane Koubaa was born in 1989 in Sfax, Tunisia. She received her BS degree in electrical engineering from the National Engineering School of Sfax (ENIS) in 2013. She is currently a Ph. D student in electrical engineering and a researcher in the Control and Energy Management Laboratory (CEMLab). Her research interests include energy management and optimization, electric vehicles, and renewable energy sources.

Lotfi Krichen was born in 1964 in Sfax, Tunisia. He received his BS degree in electrical engineering, his Ph. D degree and his university habilitation degree from the National Engineering School of Sfax (ENIS) in 1995, 1989 and 2008, respectively. He is currently a professor in electrical engineering at the National Engineering School of Sfax and the head of renewable energy systems team at the Control and Energy Management Laboratory (CEMLab). His research interests include electric machine drives, energy management, control and optimization, hybrid renewable energy sources, electric vehicles and smart grids.

N. Mekki, F. Derbel, L. Krichen and F. Strakosch
Power System State Estimation Using PMU Technology

Abstract: This paper focuses mainly on a recent method aiming to provide an optimal placement of Phasor Measurement Units (PMUs) for power system state estimation purposes. The proposed technique is generally based on synchrophasor technology which helps accurately and reliably in estimating the current status of the electrical network. In fact, establishing full system observability must consider the lack of the appropriate informations in all network nodes especially with the involvement of the existing renewable energy (RE) sources. Other investigations of FACTS systems are extended later to improve the performances and the stability of the smart grid. Experimentally, the application of efficient placement strategies will guarantee critical PMUs allocation in a 5-bus test system.

Keywords: Power systems, state estimation, optimal placement, smart grid, phasor measurement unit (PMU), FACTS.

1 Introduction

Commonly, a network constituted with generators, transmission lines and loads, and other different components requires instantaneously real-time power flow control and monitoring in order to provide mainly sustainable electricity first and balance between both demand and supply second. Hence, adopting a robust measurement infrastructure becomes extremely critical challenge aiming to have safe and effective observability of the current grid state. The major goal then is to collect and study the voltage magnitudes and phase angles at each node commonly named busbars or buses. This process, called State Estimation (SE), presents a fundamental key in the Energy management system frequently used to analyze contingencies, determine any required corrective actions and make decisions on real-time market pricing [1]. In fact, estimating the total system state in accurate way is related principally to know the voltage magnitudes and phase angles at each bus taking into account the grid topology and the different impedance parameters. Since the pioneering work of F.C. Schweppe in 1970 [2], SE has become the core stone in supervisory smart grid control and planning. Basically, being observable in a cost effective operation,

N. Mekki, F. Derbel, L. Krichen and F. Strakosch: N. Mekki, University of Sfax, Tunisia, email: mekki_nesrine@hotmail.fr, F. Derbel, Leipzig University of Applied Sciences, Germany, email:faouzi.derbel@htwk-leipzig.de, L. Krichen, University of Sfax, Tunisia, email: lotfi.krichen@enis.rnu.tn, F. Strakosch, Leipzig University of Applied Sciences, Germany, email:Florian.Strakosch@htwk-leipzig.de

De Gruyter Oldenbourg, ASSD – Advances in Systems, Signals and Devices, Volume 7, 2018, pp. 139–154.
https://doi.org/10.1515/9783110470529-009

a network needs definitely a given number of measurements including pseudo measurements and its probable redundancy at certain locations. Therefore, searching for optimal metering placements becomes seriously crucial. Literately, it is obtained through different approaches by utilizing for example the smart meter device, such as Schweppe. and al. methodology which introduces the variance's reduction of the estimated state variables. Cobelo and al. reduces mainly the estimation error for the voltage and its phase angle below a certain threshold [3]. Shaifu and al. aim to place a definite number of measurements based on a series of load flow simulations in order to drop the voltage magnitude deviation to those unmeasured busbars [3]. Likely, Singh and al. propose a procedure focusing on the error covariance matrix properties for distribution grids including distributed generation [3]. In addition to smart meters, phasor measurement units (PMUs) deployment has been developed considerably all through the network for linear SE purpose. To address expected circumstances, several methods proposed in [4] related to Optimal PMU Placement (OPP) are examined to meet target accuracy issue of the future power system, this constraint is classified into two kinds of algorithms: mathematical and heuristic algorithms. The first one was announced by Yuill and al. who provided an overview of the Integer Program and Exhaustive search [1, 5]. Likely, heuristic technique is often applied on some IEEE test networks, we outline for example the bisecting search simulated annealing, recursive security N algorithms and Spanning Tree Search [5, 6]. Mostly, the essential concern in our work looks for how guaranteeing an accurate SE in a smart grid. Thus, the work proceeds over the next steps: study of each element in a modified IEEE-14 bus system by the use of an open source software PSAT, followed by searching the optimal allocation of the appropriate measurement devices basically encountering numerous algorithms which may offer robust control and monitoring system while maintaining the cost minimal. Then, a real demonstration of theoretical results is performed via a 5-bus real plant accomplished in HTWK Laboratory, Leipzig, Germany.

2 Network development

This section focuses on the case of a modified standard IEEE 14-bus test system, it exemplifies a simple approximation of the American Electric Power system. Power flow analysis aims essentially to find the magnitude and phase angle of voltage at each bus and real/reactive power flowing in each transmission line. In fact, continuous monitoring of the actual system state and analyzing the possible grid expansions and the increased load demand are the major important targets to meet in the future network. To study the load flow analysis, PSAT affords several load level flow techniques such as Newton-Raphson [7]. The final model implemented via PSAT is displayed in Fig. 1.

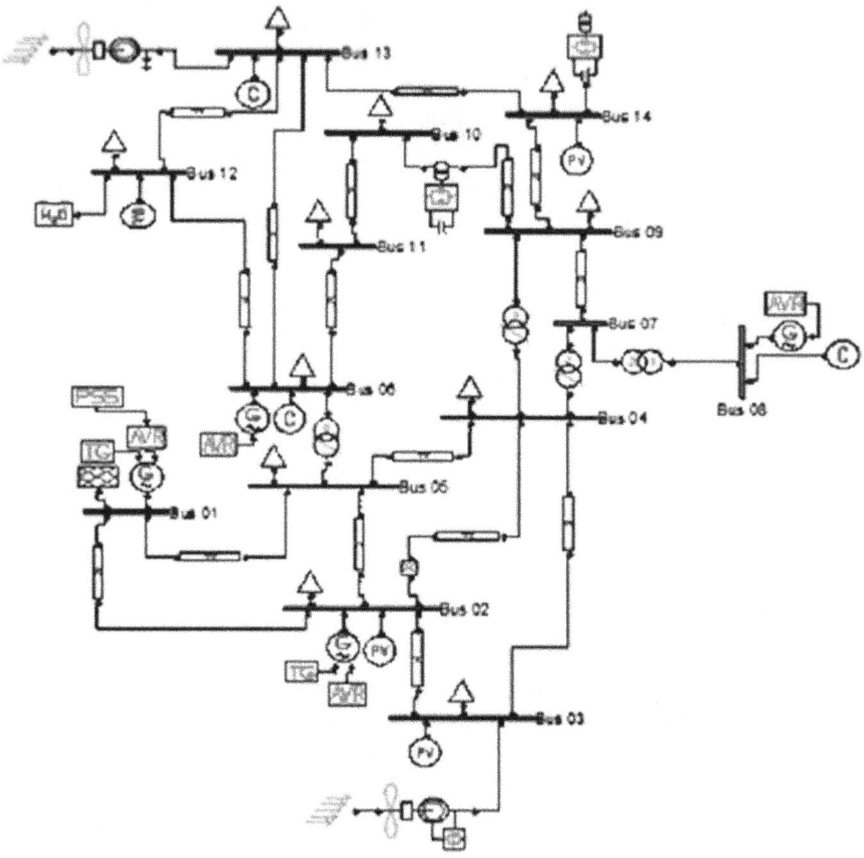

Fig. 1. Modified IEEE 14 bus system.

2.1 Grid structure

The basic aim in power system operations is to assist consumers with the needed energy and the reasonable voltage and frequency at minimum cost. In fact, utility have to take into consideration the rapid growth in load demand that forces the network components to operate near critical reactive power limits typically leading to voltage instability. The majority of plants are equipped with 3-phase synchronous generators which are implemented for voltage and reactive power control purposes. It is extremely important to keep generators functioning at synchronous speed by utilizing several regulators such as the Automatic Voltage Regulator (AVR) for voltage regulation, the Turbine Governor (TG) for turbine speed and grid frequency regulation by adjusting load fluctuations in real power and torque variabilities. Moreover, the Power System Stabilizer (PSS) when added to the AVR will help in damping out power system oscillations. To link between the different grid's parts which operate at different voltage levels, transformers are selected to be joined. Furthermore, three types of buses are examined namely the slack bus, the load bus, and the generator (PV) bus. They are usually considered as nodes which resemble numerous branch points all over the network. The load's type used in the present work is the PQ load, it is the simplest one that considers a constant demand of real and reactive powers. To transfer electrical energy from power stations to consumers, transmission lines impedance and its X/R ratio must be carried out efficiently. In fact, resistance, reactance and capacitance are three vital parameters which characterize the adopted Pi line model. Notably, when the X/R ratio increased, the voltage drop will increase as well which may affect the state variable's estimation. The grid seems to get smarter with the incorporation of RE resources, storage system, advanced metering infrastructure and several controllers. Actually, doubly Fed Induction Generators (DFIG) (connected to Bus 3) with reduced rated converters are more useful relatively than fixed speed induction generators (connected to Bus 13). Wind turbine generator features may affect the system dynamics and participate in both voltage and frequency control and deviations. Furthermore, improving the reactive power limits of the system requires urgently a remedy for prevention of voltage's instability and collapse. Analysis performed using PSAT prove that Flexible AC transmission System (FACTS) devices can be considered as efficient controllers for the growth of the system's loadability margin that is tested by utilizing the Continuation Power Flow (CPF) technique. Compared to other controllers, the most significant devices adopted in this work are the Static Synchronous Compensator (STATCOM) which is situated at the weakest bus (Bus 14) for reactive power compensation, and the Static Synchronous Series compensator (SSSC) for power flow, phase angle, and voltage magnitude control over the line causing major losses. Due to the alternating green power generation, energy storage capacity and backup generator are necessary in a smart grid configuration. The battery used is the Solid Oxide Fuel Cell (SOFC) (connected to Bus 12). It contributes in power swings damping and stability enhancement by means of its real power [8]. Adoption of particular measurement units may afford utilities with bi-directional communication systems and empowers consumers to be active participants and managers of their own energy usage. Smart meter and phasor measurement units are very essential examples of real-time and accurate metering devices which interfere in smart grid functionality and its state estimation. Thanks to its countless features and its accessibility in PSAT, adopting the PMU seems to be mostly the biggest challenge to win in smart grid topology. It is synchronized through GPS signal delivering accurate measurements of the bus voltage and current phasors via the branches incident on it. However, due to economic and technical constraints, it is impossibly feasible to place one PMU at every bus. Hereafter, the problem of optimal PMU placement will be treated in the next section. PMU computes the positive sequence of voltage and current phasors from sampled signals over the transformer secondary windings. State variable signals are oversampled to be conditioned through a combination of A/D anti-aliasing filters.

The sampling clock is phase-locked with the GPS clock pulse in order to adjust the reference frequency to the measured signal. A microprocessor at that moment estimates the phasors using basically the discrete Fourier transform and alters it to the data concentrator via the modem [1].

2.2 Renewable energy integration

In recent years, growing interests about energy cost have strengthened the concerns in harnessing energy from renewable sources to achieve the needed levels in electricity demand. Due to the RE variability, the amount of its integration into the network must be reviewed rationally. PSAT simulations can explain more its influence on the whole system as follows

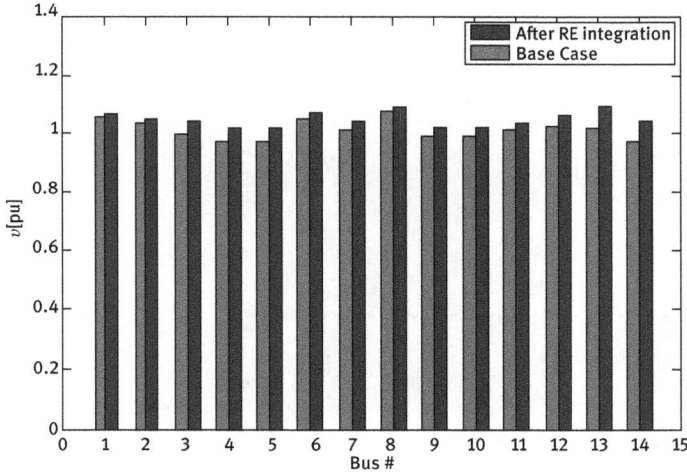

Fig. 2. Voltage magnitude profiles in the base case and after RE integration.

Figure 2 displays the voltage amplitudes of all buses in the base case (red bars), compared to RE adding case (blue bars). It demonstrates an increase of each bus voltage level just after the wind turbine integration especially in the nearest buses. Hence, RE sources incorporation should be studied and controlled carefully. Due to the fact of parallel-working generators in electrical power systems, unexpected oscillations tended to be excited, this phenomena, so-called small signal stability problem, defines the system inability to ensure synchronism once exposed to small disturbances [9]. Eigenvalue analysis shown in Fig. 3 and Fig. 4 can prove that the system have small signal stability. In other word, the more the eigenvalue is near to zero, the less the damping ratio increase.

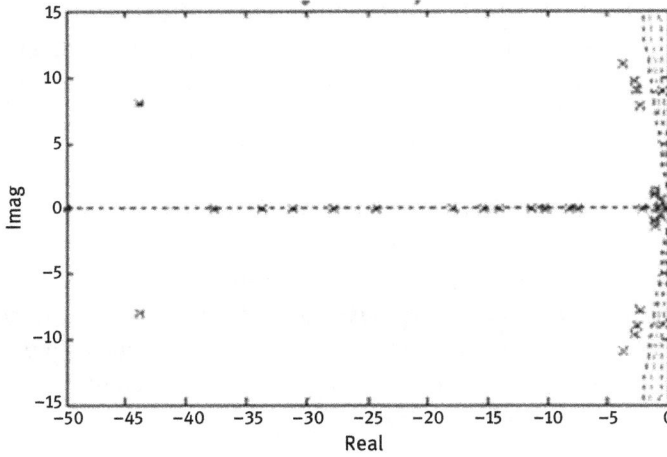

Fig. 3. Eigenvalue analysis in the base case.

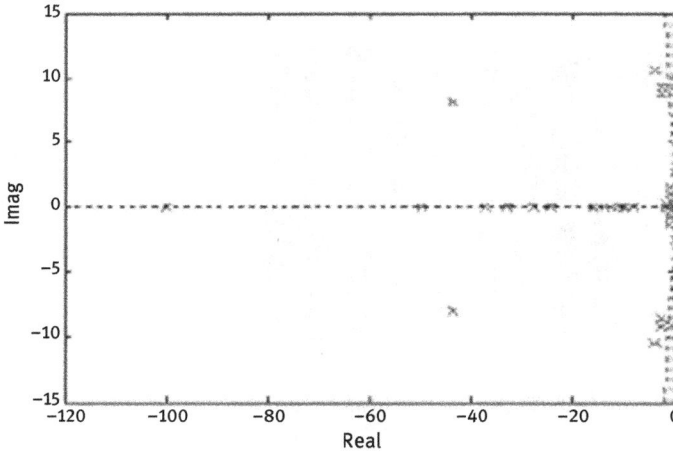

Fig. 4. Eigenvalue analysis in case of wind turbine integration.

3 Optimal PMU placement problem

Outcomes of several investigations looking at efficient OPP approaches become a challenge for the future development of powerful and well-observed energy management system. Henceforward, PMU's ability to measure line current phasors simplifies the voltage's calculation at the other end of the line using Ohm's Law. Researchers demonstrated that optimal placement of PMUs necessitates only 1/5 to 1/4 of the number of the network's buses to fulfill the desired objectives [10]. The main aim of this section is to analyze the linear static state estimation of any power system.

3.1 PMU Placement methodology

Commonly, setting the initial conditions is importantly the basis of any optimized result. They are generally selected as follows:

1. Locate a bus with the largest number of branches connected and put a PMU.
2. Identify pseudo and extended pseudo measurements.
3. Check the coverage of the whole grid, return to (1) in case of unobservability. Likewise, the illustrated rules [10, 12] for PMU placement present also an essential concept to attain the desired purposes of the present work.
   - Rule 1: Assign one voltage measurement to a bus where a PMU is placed, including one current measurement to each branch connected to the bus itself [10, 11].
   - Rule 2: Assign one voltage pseudo-measurement to each node reached by another one equipped with a PMU [10, 11].
   - Rule 3: Assign one current pseudo-measurement to each branch joining two buses where voltages are known [10, 11].
   - Rule 4: Assign one current pseudo-measurement to each branch where current can be indirectly calculated by the Kirchhoff current law (KCL). This rule is applied when the current balance at one node is identified: if the node has no power injections (if N-1 currents insert to the node are known, the last current can be computed by difference) [10, 11].

3.2 Proposed algorithm

Comparing all the suggested techniques in PSAT, the chosen optimal solution is possibly the Simulated Annealing (SA) algorithm thanks to its various features. Notably, the choice of the method is extremely related to the grid structure. In our case, practical results will confirm one more the selection of this approach which provides critically the least number of PMUs used for fulfilling full network observability. SA is a kind of heuristic random process and one of the important optimization algorithms. It is based on neighborhood search and uphill moves. It has also a strong analogy to the simulation of the material's cooling. The technique can be subdivided into a number of steps [12], as follows:

1. Generate a random solution.
2. Calculate its cost using some defined cost function
3. Generate a random neighboring solution.
4. Calculate the new solution's cost (Cnew).

5. Compare them:
   - If Cnew< Cold : move to the new solution.
   - If Cnew> Cold : maybe move to the new solution.
6. Repeat steps 3-5 above till getting an acceptable solution or when reaching the maximum number of iterations.

SA method tries mainly to find the best solution by generating a random initial solution and exploring the nearby area using hill climbing principle. If an adjacent solution is better than the current one, then it moves to it. If not, the algorithm stays put as shown in Fig. 5 and Fig. 6 [13, 14].

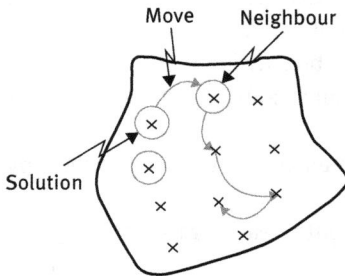

Fig. 5. Simulated annealing principle.

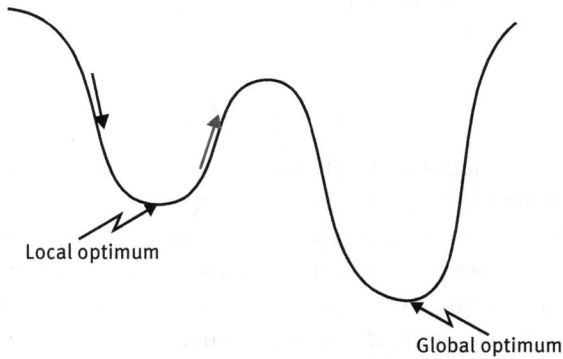

Fig. 6. Hill climbing principle.

4 Simulations and results

Owing to its high cost, it's pointless to allocate the PMU at each bus existing in the power system. Thus, optimal PMU placement should be planned as well as the network structure change. Referring to the 4 rules viewed in the previous section, only 5 PMUs are elected to be placed at different locations as represented in Fig. 7 and Fig. 8.

Particularly, bus 7 is the only zero injection bus where there is no injected power or load consumption. First, a PMU is sited at bus 4 which has the biggest number of incident branches. Its phasor measurements are apportioned to the branches 4-2, 4-3, 4-5, 4-7, and 4-9. Using Ohm's law, the voltage phasors at buses 2, 3, 5, 7, and 9 can be identified from the branch current and the nodal voltage at bus 2. Based on the known voltage phasors, the currents of the branches (pseudo-measurements) 2-3, 2-5,

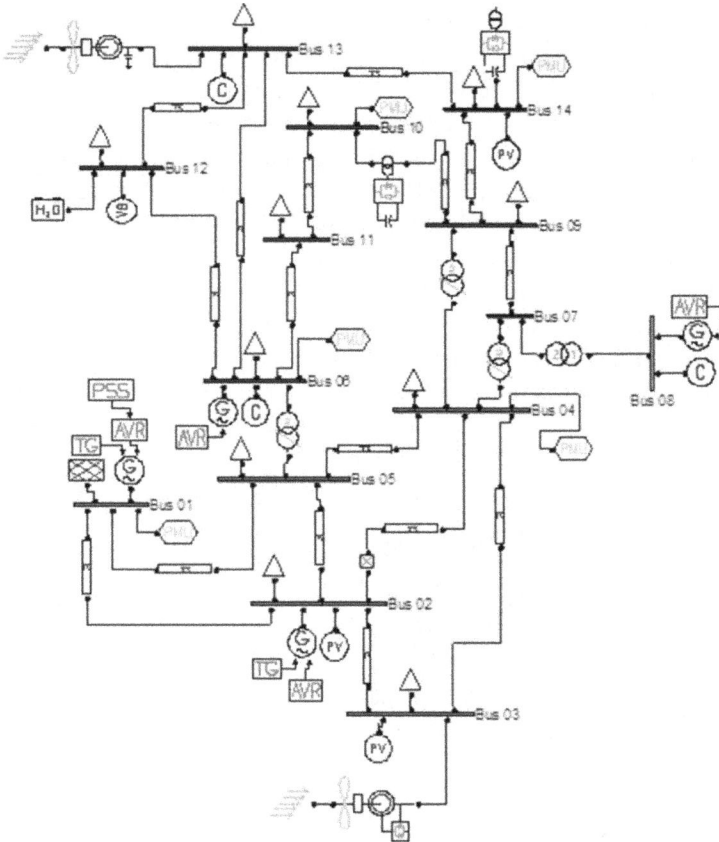

Fig. 7. PMUs placement in the grid.

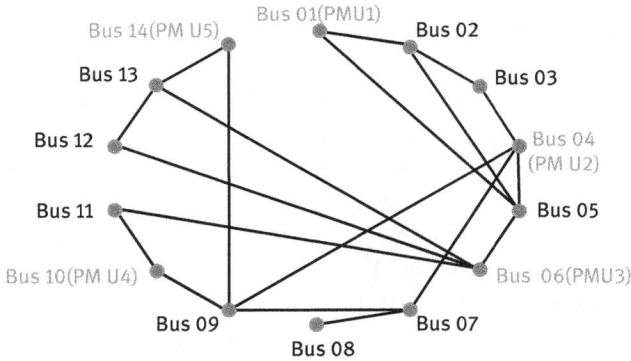

Fig. 8. Graph representation of the performed network.

and 7-9 can be gotten using Ohm's law too. Using the known currents of branches 4-7 and 7-9, the current of the branch 7-8 can be concluded using KCL. Then, the voltage phasor at bus 8 is possibly deduced by Ohm's law. As the PMU coverage at bus 4 is maximal, the technique will continue equally for the other PMUs allocations which are required at buses 1, 10, and 14, in order to make the complete system observable. Accordingly, in this example, 5 PMU's placement set is not optimal. Thus, new placements are generated by the Simulated Annealing method which ensures the system observability only with 3 PMU's at buses 2, 6, and 9 as shown in Fig. 9.

As summarized in Table 1, the direct (N-1) and direct spanning tree reveal the worst cases. The depth first and recursive (N or N-1) involve high number of PMUs which doesn't fit the major interests of the present work. For that reason, we can say that the graph theoretic method is definitely an energy effective but cost effective

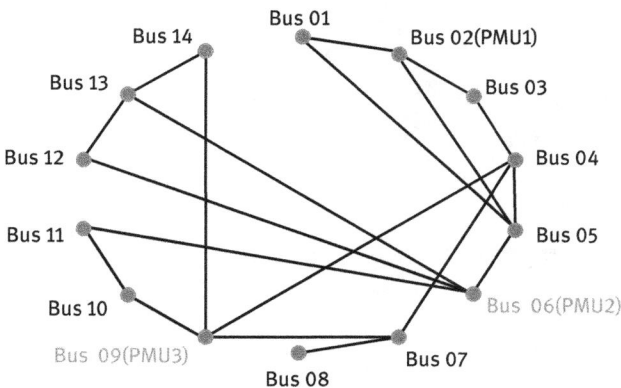

Fig. 9. Graph representation of the performed network.

Tab. 1. Simulation results for PMUs placement.

Algorithm	Number of PMUs	Number of Non Observable Buses	CPU Time (s)
Depth First	6	0	0.14127
Simulated Annealing	3	0	0.44408
Graph Theoretic Procedure	5	0	0.15581
Recursive Spanning Tree	6	0	0.50233
Direct Spanning Tree	2	11	0.55427
Recursive(N-1)Spanning Tree	8	0	0.45497
Direct(N-1)Spanning Tree	2	11	0.46133

approach. Therefore, to assure complete system observability with minimal PMUs number, simulated annealing algorithm seems to be the best one.

5 Experiment results

This section focuses on a real model accessible in the laboratories of HTWK University, Leipzig, Germany. The model is carried out using PSAT and Power Factory simulators. It is constituted by 5 buses, a synchronous generator with fluctuated power to be used virtually as a wind turbine, a feeder and a resistive load.

The objective now is to make this network observable even in case of unpredictable contingencies. In fact, the deficiency of measurement at some system points stimulates the need of allocating the PMUs at critical locations. Consequently, referring to the predefined rules of PMU placement, the system seems to require only one PMU, and the best choice is setting it at bus 4 (Fig. 10) as it has the most incident branches connected to it. To clarify how a PMU will proceed after its settlement, an example of manual computations is executed as appendix concerning two buses as shown in Fig. 11.

In general, the lows applied are:

- Ohm's low to obtain the voltage pseudo-measurement at bus 1, bus 3 and bus 6.
- Kirchhoff low to calculate the pseudo-current between buses 1 and 2 or between buses 2 and 3. In view of that, the 3 lines 'line', 'line 1' and 'line 3' have the same current.

Figure 12 shows the implemented model essentially used to study the optimal placement of the PMUs.

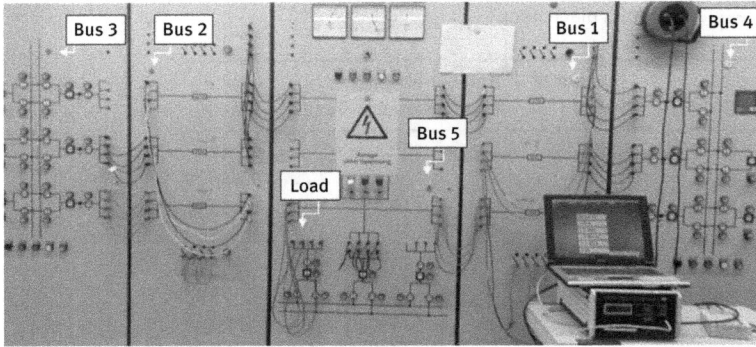

Fig. 10. Network model.

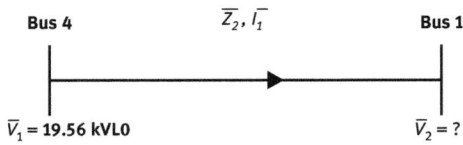

Fig. 11. Example of current and Voltage computations between Bus 4 and Bus1.

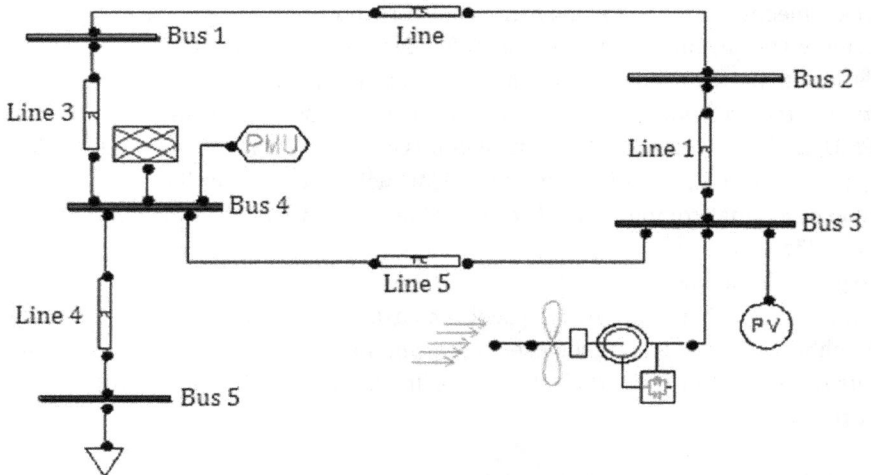

Fig. 12. PMU Settlement using PSAT.

6 Conclusion

This paper aims mainly to ensure an accurate state estimation in a smart grid sturcture in order to keep utility and consumers in touch with their instant energy usage whereas maintaining the cost minimal. This work interests first in the most important components of the implemented network, as well as the effect of renewable energy sources integration as a major concern, since it affects especially its stability, Afterwards, as the grid behavior to the successive incorporation of these components, deciding about critical PMUs placement becomes urgent in order to guarantee the entire power system observability. Therefore, different algorithms are proposed in theory and implemented in PSAT so that it will be more feasible to control fruitfully the minimum number of the state variables, and solve competently the optimal PMU placement problem at minimum cost and in a reliable way. Experimentally, due to the unavailability of PMUs in the laboratory, an example of computations was developed to demonstrate how this measurement unit can work and guarantee the whole network prevision. The future work may look possibly for the optimization of the algorithms used for sitting optimally the PMU in order to have eventually more levels of the network observability with the least state estimation error propagation. Similarly, a valuable study of excessive diffusion of renewable energy sources may be a compulsory chiefly with the incorporation of several advanced and sophisticated control devices.

7 Appendix

Example of calculations: Fig. 12

Base Data: $V_B = 20KV$, $S_B = 14.5MVA$, $Z_B = \dfrac{V_B^2}{S_B}$

For line length of 4 km (Line 3, Line 4): $\overline{Z_L} = 1.65 + j1.5$

For line length of 8 km (Line 8, Line 5): $\overline{Z_L} = 3.3 + j3$

For line length of 12 km (Line): $\overline{Z_L} = 4.95 + j4.5$

The base impedance is: $Z_B = \dfrac{20^2}{14.5} = 27.58\Omega$

The line impedance is: $\overline{Z_L} = \dfrac{1}{Z_B}(1.65 + j1.5) = 0.059 + 0.054j$

As the apparent power S_1 is defined as follows:

$$\overline{S_1} = \overline{P_1} + j\overline{Q_1} = \overline{V_1}\,\overline{I_1^\star}$$

So, we can deduce the current value after calculating the voltage value in per unit (pu)

$$\overline{V}_{1pu} = \frac{\overline{V_1}}{V_B} = \frac{19.56\angle 0}{20} = 0.978pu \quad \Longrightarrow \overline{I_1^\star} = \frac{\overline{S_1}}{\overline{V_1}}$$

Referring now to Tab. 2, the active and the reactive power are calculated in per unit as follows:

$$\overline{P}_{1pu} = \frac{\overline{P_1}}{S_B} = \frac{1.6\angle 0}{14.5} = 0.11\angle 0 \quad \Longrightarrow \overline{I_1^\star} = \frac{0.11 + j0.268}{0.978} = 0.296\angle 1.18$$

Tab. 2. Practical measurements: Case of a load = 41 Ω.

	Bus4			
	External Grid	**line 4**	**line 3**	**line 5**
U(KV)	19.56	19.27	19.54	19.61
I(A)	213	257	89	253
P(MW)	1.6	4.2	1.7	4.9
Q(MVAR)	3.9	2.7	500	1.1
S(MVA)	4.2	5.1	1.8	5
PF	0.37	0.84	0.95	0.97

$$\overline{Q}_{1pu} = \frac{\overline{Q}_1}{S_B} = \frac{3.9\angle 0}{14.5} = 0.269\angle 0 \implies \overline{I}_1 = 0.296\angle - 1.18$$

We can estimate after that the unknown voltage V_2

$$\Rightarrow \overline{V}_1 = \overline{V}_2 + \overline{Z}_L \overline{I}_1$$
$$\Rightarrow \overline{V}_2 = \overline{V}_1 - \overline{Z}_L \overline{I}_1$$
$$\Rightarrow \overline{V}_2 = 0.986\angle - 0.022$$
$$\Rightarrow V_2 = 0.986pu$$
$$\Rightarrow V_2 = 0.986 \times 20KV = 19.72KV \approx 19.54KV \text{ (Measured experimentally in Tab. 2,}$$
referring to Line 3 connecting these two buses).

Bibliography

[1] D. Echternacht, C. Linnemann and A. Moser. Optimized Positioning Of Measurements In Distribution Grids. *3rd IEEE PES Innovative Smart Grid Technologies Europe*, 2012.

[2] S. Lefebvre, J. Prevost and L. Lenoir. Distribution State Estimation: A Necessary Requirement For The Smart Grid. *IEEE PES General Meeting, Conference and Exposition*, 2014.

[3] A. Abdel-Majeed, S. Tenbohlen, D. Schollhorn and M. Braun. Meter Placement For Low Voltage System State Estimation With Distributed Generation. *22nd International Conference on Electricity Distribution, Stockholm*, 2013.

[4] B, Singh, N.K. Sharma, A.N. Tiwari, K.S. Verma and S.N. Singh. Applications Of Phasor Measurement Units (PMUs) In Electric Power System Networks Incorporated With FACTS Controllers. *International Journal of Engineering, Science and Technology*, Vol.3, No.3, pp.64–82, 2011.

[5] NM. Manousakis, GN. Korres and PS. Georgilakis. Taxonomy of PMU Placement Methodologies. *IEEE Transactions on power systems*, Vol.27, No.2, 2012.

[6] GB. Denegri, M. Invernizzi and F. Milano. *A Security Oriented Approach To PMU Positioning For Advanced Monitoring Of A Transmission Grid*. Power System Technology, 2002.

[7] B. Nitve and R. Naik. Steady State Analysis Of IEEE-6 Bus System Using PSAT Power Toolbox. *International Journal of Engineering Science and Innovative Technology (IJESIT)* Vol.3, Issue 3, 2014.

[8] A. Wank. Modification Of A Dynamic Simulation Tool For Power Networks To Analyse And Calibrate A Risk-Based OPF Approach. *Technische Universitat Munchen*, 2012.

[9] J. Machowski, JW. Bialek and JR. Bumby. *Power Sysytem Dynamics Stability And Control*. Second Edition, John Wiley and Sons, 2008.

[10] T.L. Baldwin L. Mili M.B. Boisen, Jr and R. Adapa. Power System Observability With Minimal Phasor Measurement Placement. *IEEE Transactions on Power Systems*, Vol.8, No.2, 1993.

[11] F. Milano. *Documentation for PSAT version 2.0.0*. February 14, 2008.

[12] T-T. CAI and Al. QIAN. Research of PMU Optimal Placement in Power Systems. *Int. Conf. on Systems Theory and Scientific Computation, Malta*, 2005.

[13] Petru Eles. Simulated Annealing. *Heuristic Algorithms for Combinatorial Optimization Problems*, 2010.

[14] B. Shayanfard, M. Dehghani and A. Khayatian. Optimal PMU placement for full observability and dynamic stability assessment. *Electrical Engineering (ICEE), 19th Iranian Conference*, 2011.

Biographies

Nesrine Mekki was born in 1992, Tunisia. She received the Engineer Diploma in Electrical Engineering and the Master degree in Embedded Systems, both from the National School of Engineering of Sfax in September and December 2015, respectively. Curretly, she is a PhD Student at the Control and Energy Management Laboratory (CEMLab) in the National School of Engineering of Sfax. Her research interests include Smart Grids, Power System Operation and Control, Renewable Energy Sources.

Faouzi Derbel is currently Professor for Smart Diagnostic and Online Monitoring within the Leipzig University of Applied Sciences Germany, since 2013. His research activities are focused on power aware design of wireless sensor networks and sensor systems with smart signal processing as well as state estimation of smart Grids. He received the M. Sc. degree in electrical engineering from the Technical University of Munich, Germany, in 1995 and the Ph. D. degree from the University of the Bundeswehr, Munich, in 2001. From 2000 to 2012 he was in different positions in the industrial area e. g. strategic product manager and systems engineer responsible for wireless and future technologies in fire detection systems within Siemens Building Technologies in Munich, Germany. From 2005 to 2008 he was Head of Product innovation department at Siemens Building technologies in Mühlhausen, Germany. From 2008 to 2012 he was the head of research and development department at QUNDIS Advanced Measuring Solutions, Germany. In 2012 he was professor for electrical measurement technologies at the University of Applied Sciences Zwickau, Germany. Prof.Derbel is Member of national and international technical groups, e. g. ETSI ERM TG28 dealing with short range devices within ETSI (European Telecommunication and Standardization Institute),as well as the technical group Smart Cities within DKE (Deutsche Kommission Elektrotechnik Elektronik Informationstechnik im DIN und VDE). He is editor in Chief of the ASSD (Advances in Systems, Signalsand Devices, Issues on Communication and Signal Processing and Power Systems and Smart Energies, De-Gruyter Verlag, Germany) and holds many awards and patents.

Lotfi Krichen was born in Sfax, Tunisia. He received the Engineer Diploma, the Doctorate degree and the University Habilitation in Electrical Engineering, all from the National School of Engineering of Sfax in 1989, 1995 and 2008, respectively. Currently, he is a Professor of Electric Machines and Renewable Energies in the Electrical Engineering Department of the National School of Engineering of Sfax. His research interests are Motor Drives, Energy management, Renewable Energy Systems, Hybrid Electric Vehicles and Fuel Cell Systems.

Florian Strakosch graduated as B. Eng. in automation technology and information systems and received his M. Sc. in electrical engineering at the Hochschule für Technik, Wirtschaft und Kultur Leipzig, Germany, in 2010 and 2012, respectively. He is currently working towards the Ph. D. degree in the area of wireless sensor networks for industrial environments. His main fields of research are robust energy harvesting strategies, energy aware RF communication and nonlinear forecasting algorithms for state estimation.

T. Bakir, B. Boumedyen, P.F. Odghaard, M.N. Abdelkrim and C. Aubrun

A New Wavelet–ANN Approach Based on Feature Extraction for a FAST Wind Turbine Model Diagnosis System

Abstract: This work presents a method to increased the fault detection accuracy in wind turbine system using a combination of Wavelet, Principal Component Analysis (PCA), Parseval's theorem and Neural Networks. Preprocessing, feature extraction and classification rules are three crucial issues for fault detection. To employ this issues, a two stages hybrid approach is used. The first stage is composed of preprocessing and feature extraction steps, where wavelet transform is exploited for preprocessing residual signals while two tools are used and compared to extract features based on PCA and Parseval's theorem. During the classification stage, the Artificial Neural Network (ANN) is explored to achieve a robust decision in presence and absence of faults. The approaches are applied on a FAST wind turbine system.

Keywords: FAST Wind turbine system, fault detection and isolation, Wavelet, Artificial Neural Network, Parseval's theorem, Principal Component Analysis

1 Introduction

Over the past few years, renewable energy or nonconventional energy sources is increasingly important and have acquired much attention due to the late energy crisis and the benefit to get clean energy. It is a strong contender because of its relative cost competitiveness, reliability due to the maturity of the technology and good infrastructure. In the last decade, wind energy has established itself as a significant provider of electrical power in the world and the size of wind turbines has become physically larger in order to harvest wind energy more efficiently.

Depending on the blade pitch angles, the rotor speed and the wind speed, a torque will be developed by the rotor. This torque acts on the main shaft of the turbine

T. Bakir, B. Boumedyen, P.F. Odghaard, M.N. Abdelkrim and C. Aubrun: T. Bakir, University of Gabes, Tunisia, email: tahani.bakirt@gmail.com, B. Boumedyen, University of Gabes, Tunisia email: boumedyen.boussaid@gmail.com, P. F. Odghaard, Aalborg University, Danemark, email: odgaard@ieee.org, M.N. Abdelkrim, University of Gabes, Tunisia email: naceur.abdelkrim@enig.rnu.tn, C. Aubrun, Centre de Recherche en Automatique de Nancy, Lorraine University, BP 70239 - 54506 Vandoeuvre Cedex, France, email: christophe.aubrun@univ-lorraine.fr

De Gruyter Oldenbourg, ASSD – Advances in Systems, Signals and Devices, Volume 7, 2018, pp. 155–178.
https://doi.org/10.1515/9783110470529-010

and, depending on the counter-torque exerted by the generator, the drive train will accelerate. This mode is considered essentially as a rigid-body mode and hence if the torque balance is constant, the drive train will continue to accelerate or decelerate. A wind turbine left on its own is a faulty system.

Thus, repair, diagnosis and maintenance are difficult. That's why, it is a crucial task to reduce the operational and maintenance costs to make wind turbines as competitive as the classical electric power stations. The most efficient way to improve safety considerations, to minimize maintenance costs, to lower the frequency of sudden breakdowns and to provide reliable power generation should be to monitor the condition of these systems continuously. Wind turbines must be monitored from time to time to ensure that they are in good condition.

A deep knowledge about all the phenomena concerned during the occurrence of a failure constitutes an essential background for the development of any failure diagnosis system. Many techniques and tools are available for the condition monitoring of wind turbines in order to extend their life span [1, 2]. Thanks to advances in signal processing technology, it is now possible to utilize wavelet principles to efficiently diagnose. Recently, applications of the wavelet transform (WT) and artificial neural network (ANN) in diagnosis fields [3] are various and can be presented in several studies that refer primarily on the signal processing and classification in different area [3, 4]. Recent advances in the field of neural networks have made them interesting for analyzing signals. So that, they have been successfully used in a diversity of diagnosis applications in wind turbine system [5, 6]. ANNs not only model the signal, but also make a decision as to the class of signal.

The authors in [7], [8] and [9] demonstrated the advantage of features extraction from the wavelet transform coefficients at different scales as inputs to neural networks for classifying the non-stationary signal type.

Integrating the DWT methodology with the artificial intelligence method or expert system to become a practical power disturbance classifier for recognizing accurately the disturbance has attracted much research interest [10]. However, two practical problems must be overcome in the above methods. Firstly, adopting directly the DWT coefficients as inputs to the neural networks requires large memory space and much learning time. Second, The decomposition level with the number of extraction features must be reduced to enhance computing efficiency and accuracy of recognizing the disturbance type.

In this work, which is an extension of our last work [11] the original residual signals using WT is decomposed into approximations and details versions with different frequency bands by using a successive low-pass and high pass filtering. The decomposed levels will not change their information in the time domain. However, useful information can be contained in some sub-bands. So, the fault can be detected from a given level of resolution. This is based on a choice of an indicator to determine the optimal levels where failure can occur. The residual signals are decomposed using the Daubechies wavelet. Clearly, useful information is contained in some decomposition levels. In order to extract useful information, the energy distribution is established by Parseval's theorem and PCA tools. The latter methods are used as principal criterion to select the optimal level of resolution. A comparison between the two tools will be given and interpreted.

The Parseval's theorem is used to calculate the main details which contain more information about the residual signal. This theorem is also known as the energy theorem [12, 13].

Principal Component Analysis (PCA) is a dimensionality reduction technique without much loss of information [14]. Though the main idea of PCA appeared in 1889 and it's research and applications are still in study. It is a statistical technique that is greatly used in signal processing and also used for finding pattern in high dimension data. The principle components are determined using the raw data. It produces lower dimensional data representations, thus eliminating the noise effects, and therefore increasing the efficiency of fault detection and diagnosis. In an other hand, the structure derived from PCA is helpful in identifying the variables responsible for the fault [15].

The remainder of this paper is structured as follows. Section 2 describes the fault diagnosis method. It describes the wavelet transform, the Parseval's theorem, the PCA analysis used for generating inputs to the artificial neural network (ANN) for fault classification. Section 3 presents the case study of the FAST wind turbine. Results and discussion are presented in Section 4. Finally, the main conclusions are outlined in Section 5.

2 Fault diagnosis method

In this section, a diagnosis method, which consists of two approaches, namely, the WT and Parseval's theorem, and the second one is WT and the PCA analysis are described to monitor the wind turbine system.

2.1 Basics of the discret wavelet transform (DWT) and the multi-resolution analysis (MRA)

The wavelet analysis block transforms the signal into different time-frequency scales [16]. Using DWT, a signal can be broken down into some signals in distinct frequency bands, which are called wavelet coefficients. It is more suitable for study of transient states in comparison with other methods such as Fourier transform (FT). It breaks down a signal to an approximate and a detail. The wavelet transform (WT) uses two functions which are the wavelet function φ and scaling function ϕ to perform simultaneously the multi-resolution analysis (MRA) decomposition and reconstruction of the measured signal.

The approximate signal is again broken down in order to generate another step, and this process is repeated. In fact, the original signal, by passing through 2 high pass and low pass filters, will be decomposed into 2 signals as approximate and detail, which are shown in figure 1.

The wavelet φ function will generate the detailed version (high-frequency components) of the decomposed signal and the scaling function ϕ will generate the approximated version (low-frequency components) of the decomposed signal.

In this figure, d_i and a_i signals include high and low frequencies of the main signal, respectively. In the next step, this breaking is continued with the breaking down of signal. The decomposition starts by passing a signal through these filters. The coefficients a_j and d_j are computed using the tree decomposition algorithm allowing storing low frequency information of the signal as well as the discontinuities. Finally, the signal is decomposed at the expected level.

In general, the discrete φ and ϕ can be defined as follows [16]:

$$\varphi_{j,n}[t] = 2^{\frac{i}{2}} \sum_n d_{j,n} \varphi \left[2^j t - n \right] \tag{1}$$

$$\phi_{j,n}[t] = 2^{\frac{i}{2}} \sum_n a_{j,n} \phi \left[2^j t - n \right] \tag{2}$$

where a_j and d_j are simultaneously the scaling coefficient at scale j and the wavelet coefficient at scale j.

The two functions must be orthogonal and satisfy the properties as follows:

$$\begin{cases} \langle \phi \cdot \phi \rangle = \frac{1}{2^j} \\ \langle \varphi \cdot \varphi \rangle = \frac{1}{2^j} \\ \langle \varphi \cdot \phi \rangle = 0 \end{cases} \tag{3}$$

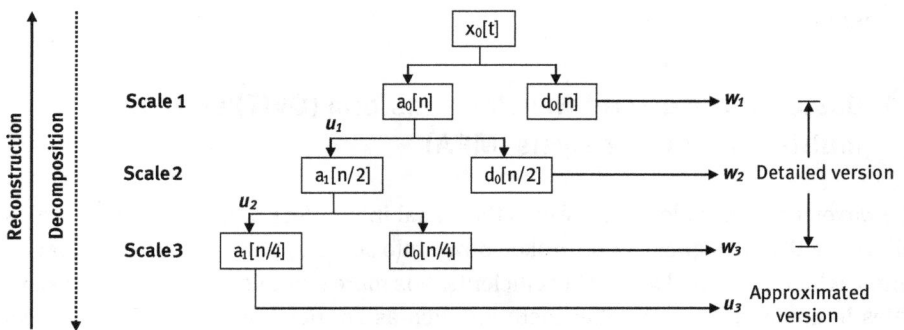

Fig. 1. Multi-resolution analysis (MRA) decomposition.

Let consider the original signal $x_j[t]$ at scale j is sampled at constant time intervals, then $x_j[t] = (v_0, v_1, ..., v_{N-1})$, the sampling number $N = 2^J$ and J is an integer number. The DWT mathematical recursive equation for a signal $x_j[t]$ is given by the following equation:

$$DWT(x_j[t]) = \sum_k x_j[t]\,\phi_{j,k}[t]$$
$$= 2^{\frac{(J+1)}{2}}\left(\sum_n u_{j+1,n}[t]\,\phi\left[2^{j+1}t-n\right] + \sum_n w_{j+1,n}[t]\,\varphi\left[2^{j+1}t-n\right]\right) \qquad (4)$$
$$0 \le n \le \frac{N}{2^j} - 1$$

Where

$$u_{j+1,n} = \sum_k a_{j,k} v_{j,k+2n} \quad 0 \le k \le \frac{N}{2^j} - 1 \qquad (5)$$

$$w_{j+1,n} = \sum_k d_{j,k} v_{j,k+2n} \quad 0 \le k \le \frac{N}{2^j} - 1 \qquad (6)$$

$$d_k = (-1)^k c_{2p-1-k}, \qquad p = \frac{N}{2^j} \qquad (7)$$

where $u_{j+1,n}$ is the is the approximated version at scale $j+1$ and $w_{j+1,n}$ is the detailed version at scale $j+1$. The signal can be reconstructed using both $u_{j+1,n}$ and $w_{j+1,n}$ coefficients. This is called the inverse discrete wavelet transform (IDWT). Figure 1 illustrates the three decomposed/reconstructed levels of the DWT algorithm.

2.2 Feature extraction and feature selection

Feature extraction is the unique process that transforms the raw signals from its original form to a new form to extract suitable information from them. The feature extraction step is crucial in an automatic classification system. This is because a classifier can operate reliably only if the features of each event are selected properly. During fault, the amplitude and frequency of the test signal will change significantly as the system change from normal state to faulty state. Two typical principles of signal classification based on feature extraction are presented in this work. The first deals with the Parseval's energy. According to [13], the energy of the distorted signal can be divided at different resolution scales in different scenario depending on the residue signal problem. The objective of the proposed method is to demonstrate the effectiveness of the energy distribution as principal criterion for selecting the optimal decomposition level. The level having the largest value indicates the desired level. The second method is based on the Principal Component Analysis (PCA).

2.3 First method of feature extraction: Parseval's theorem

The first method consist on calculating with the distribution of energy in details. Energy of the details given from the DWT stage is calculated using Parseval's theorem. This theorem refers that the sum of the square of a function is equal to the sum of the square of its transform. Using wavelet coefficients Parseval's theorem can be stated as the energy that a time domain function contains is equal to the sum of all energy concentrated in the different decomposition levels of the corresponding wavelet transformed signal. According to the Parseval's theorem, the consumptive resistance's energy of a discrete signal $x[n]$ is the square sum of the spectrum coefficients of the Fourier transform in the frequency domain:

$$\frac{1}{N} \sum_{n=\langle N \rangle} |x[n]|^2 = \sum_{n=\langle N \rangle} |a_k|^2 \tag{8}$$

where N is the sampling period, and a_k is the spectrum coefficients of the Fourier transform. In our case, we will apply the parseval's theorem to the DWT. Using 4 and 8, we obtain the Parseval's theorem in the DWT application:

$$\frac{1}{N} \sum_t |x[t]|^2 = \frac{1}{N_J} \sum_k |u_{J,k}|^2 + \sum_{j=1}^{J} \left(\frac{1}{N_j} \sum_k |w_{j,k}|^2 \right) \tag{9}$$

therefrom, through the DWT decomposition, equation 9 shows the energy of the distorted signal. The first term on the right of 9 concerned the average power of the approximated version of the decomposed signal, while the second term denotes that of the detailed version of the decomposed signal. The second term presenting the energy distribution features of the detailed version of distorted signal will be employed to extract the features of residue signals.

2.4 Second method of feature extraction: PCA

The second method consist on calculating with the principal component of details. Principal components of the details given from the DWT stage is calculated using PCA analysis. Principal Component Analysis is a technique which uses sophisticated underlying mathematical principles to transform an amount of possibly correlated variables into a smaller number of variables which are called principal components. Its aim is to reduce the large dimensionality of the observed variables (data space) to the smaller intrinsic dimensionality of independent variables (feature space), which are required to describe the data economically [17]. This is the case when there is a strong correlation between observed variables. It is a statistical method under the common title of factor analysis which can be used for prediction, redundancy removal, feature extraction, data compression, etc.

Principal components are particular linear combinations of the p random variables $X_1, X_2, \ldots, X_p$ with three important properties:

1. The principal components are uncorrelated,
2. The first principal component has the highest variance, the second principal component has the second highest variance, and so on,
3. The total variation in all the principal components combined is equal to the total variation in the original variables $X_1, X_2, \ldots, X_p$. They are easily obtained from an eigne analysis of the covariance matrix or the correlation matrix of $X_1, X_2, \ldots, X_p$

Note that principal components from the covariance matrix and the correlation matrix are usually not the same. In addition, they are not simple functions of the others. When some variables are in a much bigger magnitude than others, they will get heavy weights in the leading principal components. That is why, if the variables are measured on scales with extensively different ranges or if the units of measurement are not commensurate, it is better to perform PCA on the correlation matrix.

The most common derivation of PCA is in terms of a standardised linear projection which maximizes the variance in the projected space. For a given p-dimensional data set Ξ, the m principal axes $T_1, T_2, \ldots, T_m$ where $1 \leq m \leq p$, are orthogonal axes onto which the retained variance is maximum in the projected space. In general, $T_1, T_2, \ldots, T_m$ are given by the m leading eigenvectors of the sample covariance matrix given by:

$$S = \frac{1}{N} \sum_{i=1}^{N} (x_i - \mu)^T (x_i - \mu) \tag{10}$$

where $x_i \in \Xi$, μ is the sample mean and N is the number of samples. We have now,

$$ST_i = \lambda_i T_i, \quad i \in 1, \ldots, m \tag{11}$$

where where λ_i is the i th largest eigenvalue of S. The m principal components of a given observation vector $x_i \in \Xi$, μ are given by $y = [y_1, \ldots, y_m] = \left[T_1^T x, \ldots, T_m^T x \right] = T^T x$

The m principal components of x are uncorrelated in the projected space.

2.5 Feature selection

In any classification or recognition problem, feature selection is an important algorithm to identify those features with relatively small intra-class and large inter-class variations. Since such features are more discriminative, they result in more accurate classification. Furthermore, a smaller number of features reduces the computational burden of the classification process. In feature extraction, the processing of signals is necessary for the awareness of different contexts of these signals in different domains. In this section, we describe the algorithm used to select features (details from the DWT and MRA step) to be the ANN inputs. In the two tools of feature extraction, there exist details which contain more information than others. In figures 2 and 3, the flowcharts of the proposed algorithms are presented for the two tools.

2.6 Classification of residue signals by artificial neural network

Methods of distinguishing patterns have been built on the basis of many mathematical approaches and classification of information is carried out on the basis of previous experiences or statistical information of patterns. The classifying consists of three main sets of information:

1. Training information set, including classes and corresponding features.
2. Test information set, testing accuracy of classifying.
3. Validation information set.

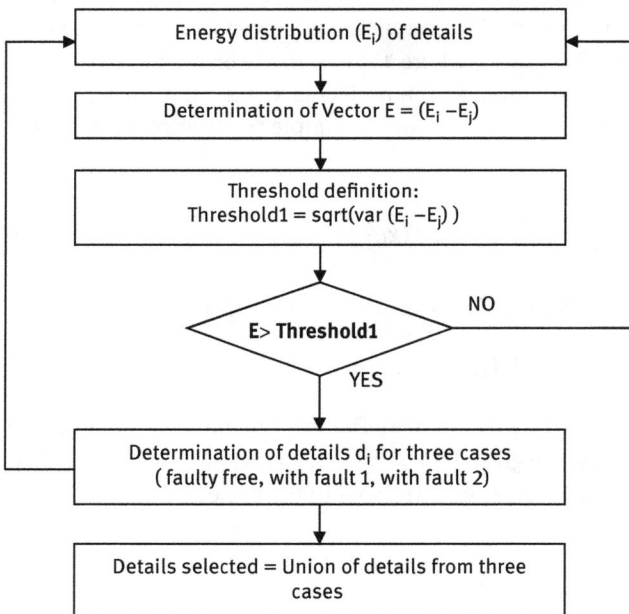

Fig. 2. Algorithme diagram for feature selection of details for the method based on Parseval's energy.

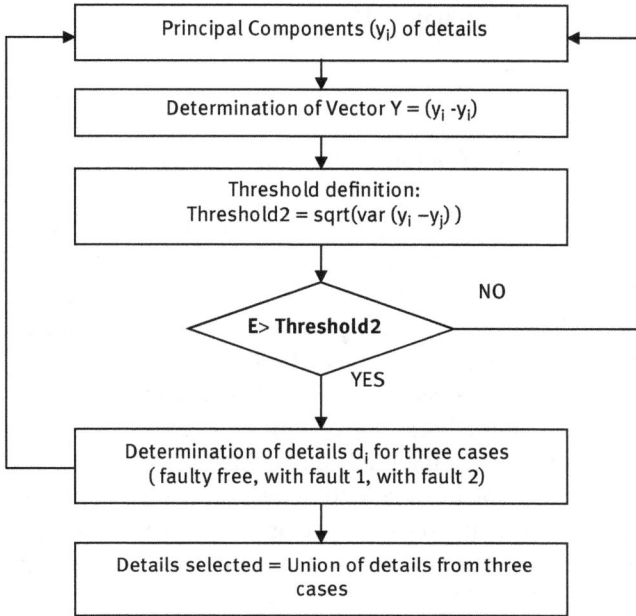

Fig. 3. Algorithme diagram for feature selection of details for the method based on PCA.

Over the past decades, serious attempts have been made in order to simulate biological neural networks, which led to introducing ANN.

In the simplest form of a layered artificial neural network, single layer feed-forward network, there is an input layer of source nodes which projects onto an output layer of neurons (computation nodes), yet not vice versa. The network is enabled to extract higher order statistics if there exist one or more hidden layers, whose computation nodes are correspondingly called hidden neurons. In the hidden layer, the activation function intervene between the external input and the network output. It is goal is to train on the basis of presented patterns which present the main capability of an ANN. The output of the artificial neural network is obtained from the equation 12.

$$y = f(w, x) = \sum_{i=1}^{n} w_i x_i \tag{12}$$

where x_i and y are the ith input and output, respectively. By optimizing, in the training steps, all variables (w_i), ANN reaches the closest answer. The sum of error squares will reach the least adequate level in terms of equation 13.

$$E_k = \sum_k \frac{1}{2}(t_{kj} - o_{kj})^2 \tag{13}$$

where t, o, and j are target, real value of output, jth output neuron, and training inputs, respectively.

The artificial neural network type used in this article is a multi-layer feed-forward network. This type of neural network is commonly referred to as multi-layer perceptron (MLP). The general architecture of a neural network MLP is shown in figure 4. MLP have been applied successfully to solve some hard and diverse problems by training them with a highly popular algorithm called the error back-propagation algorithm. These networks have been successfully used in the solution of difficult pattern- recognition problems. Standard back-propagation is a gradient descent algorithm that minimises the error between the predicted and actual output values. After each training cycle, an adjustment of the weighted connections between neurons takes place until the error in the validation data set begins to rise. The data set of validation is a second data set that is given to the network to evaluate during training. Note that if this approach is not used, the network will represent the training data set too well and will then be unable to generalize to an unseen data set or a testing data set. Once the networks are trained to satisfaction, it can be operate when the new input data are given to the the trained network in its non-training mode to produce the desired outputs. To validate the performance of the trained network before it is used into real operation, however, the operation mode is mostly initiated by utilizing the test data set. An crucial way to help promote generalisation to unseen data is to

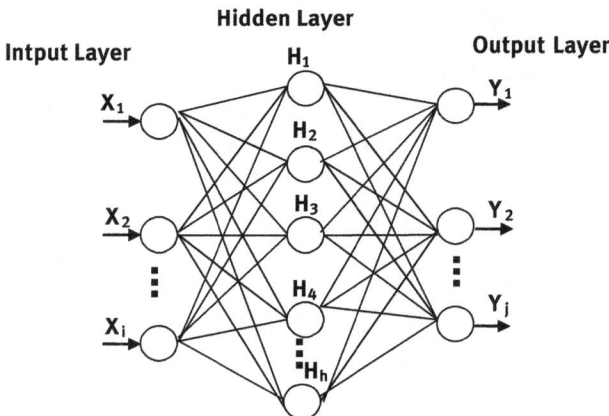

Fig. 4. General architecture of a neural network MLP.

guarantee that the training data set contains a representative sample of all the attitude in the data. This could be achieved by ensuring that all three data sets (training, validation and test) have similar properties.

In our case, Levenberg-Marquardt (LMA) is the training algorithm used which is an adaptive back propagation (BP) algorithm. It'is considered in two steps [18]. In the first step, the inputs are introduced to the network, which propagate forward to generate the output for each neuron, $y_j(t)$, in the output layer. $y_j = f(z_j)$ determines the activity of each neuron, where $f(z) = 1/(1 + e^{-z})$ the sigmoid activation function.

3 Case study

3.1 FAST wind turbine model

The FAST code [19],[20] is an aerolastic simulator capable of predicting the extreme and fatigue loads of two and three bladed Horizontal Axis Wind Turbines (HAWTs). The National Renewable Energy Laboratory (NREL) and its academic and industry partners have created this aeroelastic simulators. FAST model simulator has been accepted by the scientific community and is used by many researchers in the development of new control systems for wind turbines [21, 22]. We select this simulator for validation due to the fact that in 2005 the Germanischer Lloyd WindEnergie evaluated FAST and found it suitable for the calculation of onshore wind turbine loads for design and certification. Numerical validation with FAST were performed with the NREL 5MW on-shore wind turbine [19], [20]. The wind turbine characteristics are summarized in Table 1, [11, 20]. Table 1 presents the additional gross properties for the NREL 5-MW baseline wind turbine model.

The wind data is generated with a stochastic, full-field, turbulent-wind simulator developed by NREL. In this model, we note that the mean wind speed at the hub

Tab. 1. Gross properties chosen for the NREL 5-MW baseline wind turbine [20].

Rating	5 MW
Rotor Orientation, Configuration	Upwind, 3 Blades
Control	Variable Speed, Collective Pitch
Drive train	High Speed, Multiple-Stage Gearbox
Rotor, Hub Diameter	126 m, 3 m
Hub Height	90 m
Cut-In, Rated Cut-Out, Wind Speed	3 m/s, 11.4 m/s, 25 m/s
Cut-In, Rated Rotor Speed	6.9 rpm, 12.1 rpm
CRated Tip Speed	80 m/s

height is 18.5 m/s, thus the wind turbine is working in the full load region (also called Region 3).

The aerodynamic of the wind turbine is represented by [22]as follow:

$$\tau_r(t) = \frac{\rho \pi R^3 \, C_q[\lambda(t), \beta(t)] \, v(t)^3}{2 \, \omega_r(t)} \tag{14}$$

where C_q is the torque coefficient, ρ is the air density, R is the radius of blades, v is the wind speed and β is the pitch angle. It should be noted that the estimation of the $\tau_r(t)$ is based on measured $\beta(t)$ and $\omega_r(t)$ and an estimated $v(t)$.
$\lambda(t)$ is the tip speed ratio defined as:

$$\lambda(t) = \frac{\omega_r(t) \, R}{v(t)} \tag{15}$$

The model of the drive train is given by [22]:

$$\dot{\omega}_r(t) = \frac{1}{J} \left(\tau_r(t) - \tau_g(t) \right) \tag{16}$$

with

$$\dot{\tau}_g(t) = p_{gen}(\tau_{ref}(t) - \tau_g(t)) \tag{17}$$

where $\tau_g(t)$ is the generator torque and $\tau_{ref}(t)$ is its reference. p_g is the generator power. The wind turbine model description in the continuous-time domain is.:

$$\begin{cases} \dot{x}_c(t) = f_c(x_c(t), u(t)) \\ y(t) = x_c(t) \end{cases} \tag{18}$$

where $u(t) = [\tau_{ref}(t), \tau_r(t)]^T$ and $y(t) = x_c(t) = [\omega_r(t), \tau_g(t)]^T$. $f_c(.)$ represents the continuous-time nonlinear function that will be subsequently approximated via discrete-time fuzzy prototype from N sampled data $u(t)$ and $y(t)$, with $t = 1, 2, ..., N$. Lastly, the model parameters and the map $C_p(\lambda, \beta)$ are chosen as given in [22] in order to represent a realistic turbine.

A model-based approach is used to estimate the outputs of the system from a data driven [23],[24]. For the residue generation, sensor faults of the system under diagnosis are treated based on the data driven measurement for uncertain sequences $u(t)$ and $y(t)$.

In this work, we consider a techniques based on fuzzy logic for modeling the nonlinear processes [23, 25], [24]. The fuzzy identification and modeling is based on 'Takagi-Sugeno' (TS) model [25]. The residues r expression is as fellow:

$$r(t) = \hat{y}(t) - y(t) \tag{19}$$

where y and $\hat{y}$ are respectively the output data driven acquired from the system and its estimated given by the block of fuzzy identification. The faulty generated signal of the residue will be treated with a suitable wavelet and artificial neural network in order to detect faults.

Faults can corrupt both actuators and sensors. As it has been described in [21], sensors faults include measurements that are offset from the true values, stuck, scaled from the true values. In the table 2, faults scenarios are summarized.

3.2 Simulation results and interpretation

In this section, the result of the proposed approach is presented.An scheme block diagram for classification of residue signals is presented on figure 5.

Using the method of generation residue signals described in [23, 24, 26]. Let consider the system 18. The output of the system considered as the rotor speed angular is affected by faults scenarios described in table 2. In our case, we consider only fault 5 from table 2 and fault 9 for the two scenarios faulty cases.

The proposed method is applied to the diagnosis of the rotor speed angular. The multiresolution analysis is applied by using the Daubechies wavelet of order 4 (db4). 12 levels of decomposition are considered [23].

The input of the classification step is a preprocessed signal. In this case, residue signals in the time domain are transformed into the wavelet domain before applying as input to the neural networks. The energy distribution for the three cases respectively of each level is shown in figures 6, 7, 8. The decomposition levels 1 to 12 represent the detailed version. The figures show the obvious difference between levels.

The principal components for the three cases respectively of each level is shown in figures 9, 10, 11. The decomposition levels 1 to 12 represent the detailed version. The figures show the obvious difference between levels.

Feature extraction is the key for classification which is based on the Parseval's theorem and PCA analysis. The objective of the proposed method is compare between

Tab. 2. Faults scenarios.

Fault	Type	Time (sec)
Blade root bending moment sensor	Scaling	20–45
Accelerometer	Offset	75–100
Generator speed sensor	Scaling	130–155
Pitch angle sensor	Stuck	185–210
Generator power sensor	Scaling	224–265
Low speed shaft position encoder	Bit error	225–320
Pitch actuator	Abrupt change in dynamics	370–390
Pitch actuator	Slow change in dynamics	440–465
Torque offset	Offset	495–520
Yaw drive	Stuck drive	550–575

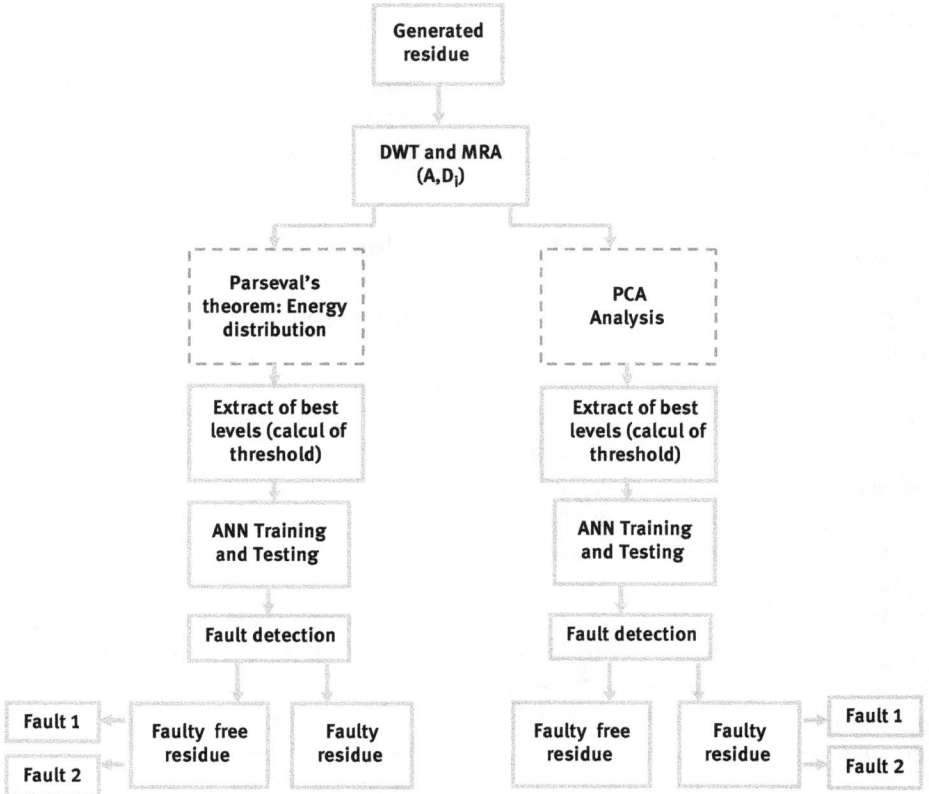

Fig. 5. Algorithm diagram.

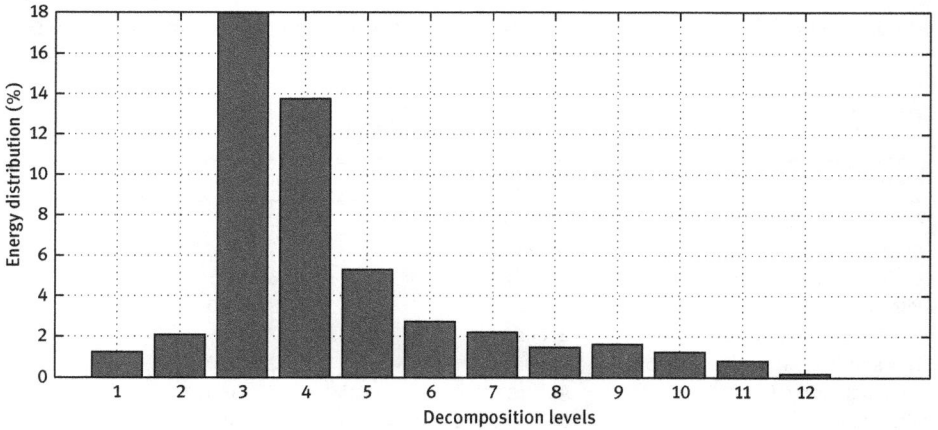

Fig. 6. Energy distribution: Faulty free residue signal.

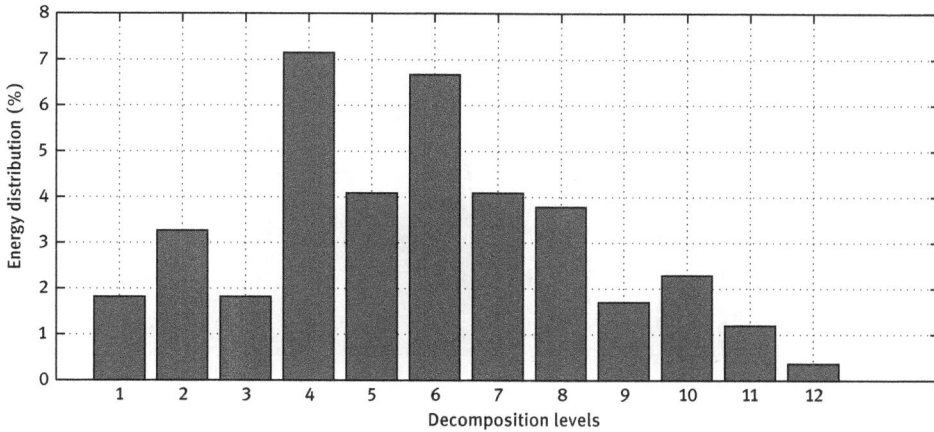

Fig. 7. Energy distribution: Fault 1.

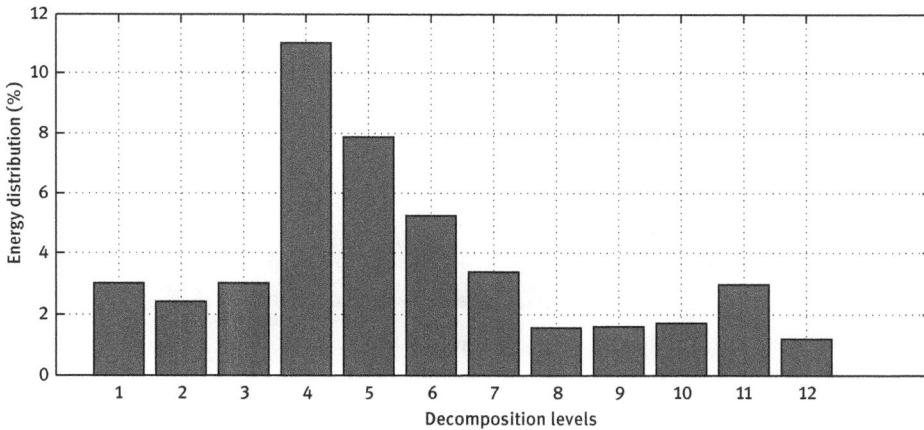

Fig. 8. Energy distribution: Fault 2.

the effectiveness of the energy distribution and PCA analysis as principal criterion for selecting the optimal decomposition level. The level having the largest value indicates the desired level. According to the proposed method described in section 2 in the subsection of feature selection, figures 12 and 13 illustrate the variance and the thresholds method in the two proposed approaches. The figures present the three classes of residual signals where class 1 is for faulty free residue (a), class 2 concerns residue with fault 1 (b) and class 3 is consecrated for residue with fault 2 (c).

The details selected from 12 details are only 4 details. Details 3,4,5,6 and Details 2,3,5,8 are respectively the selected details for Parseval's theorem approach and PCA approach.

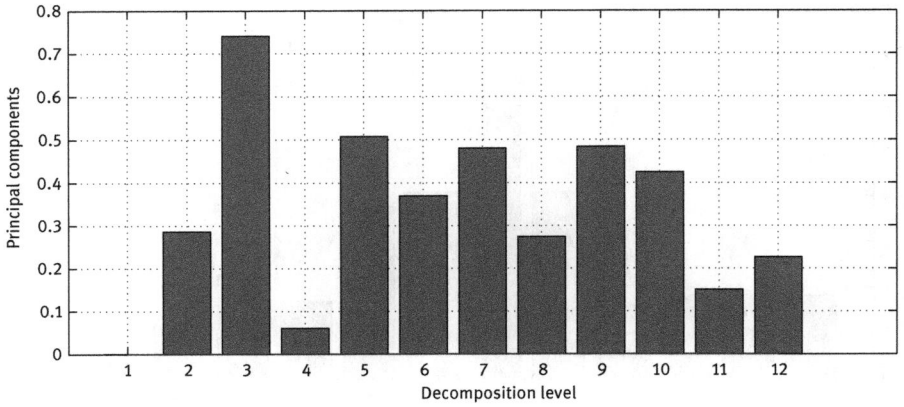

Fig. 9. Principal components: Faulty free residual signal.

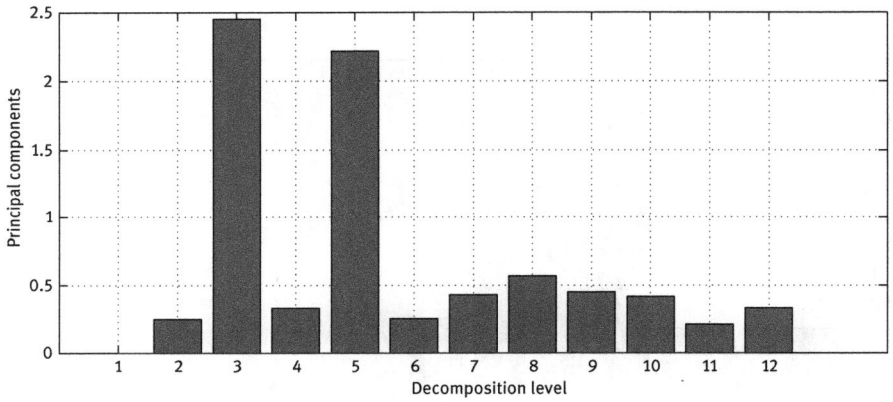

Fig. 10. Principal components: Fault 1.

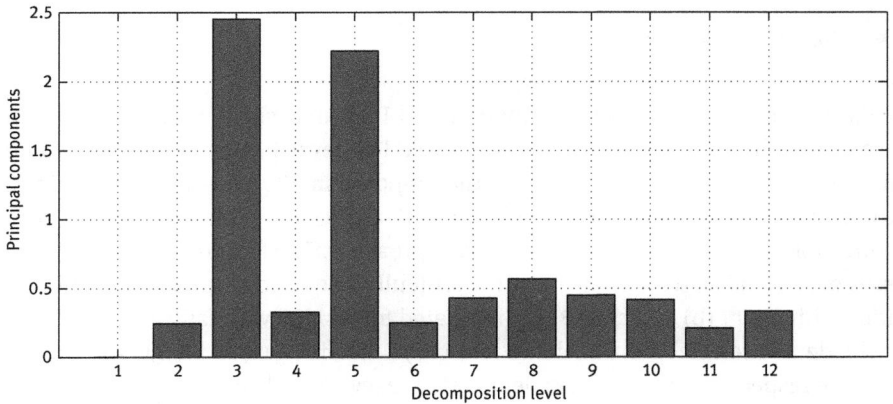

Fig. 11. Principal components: Fault 2.

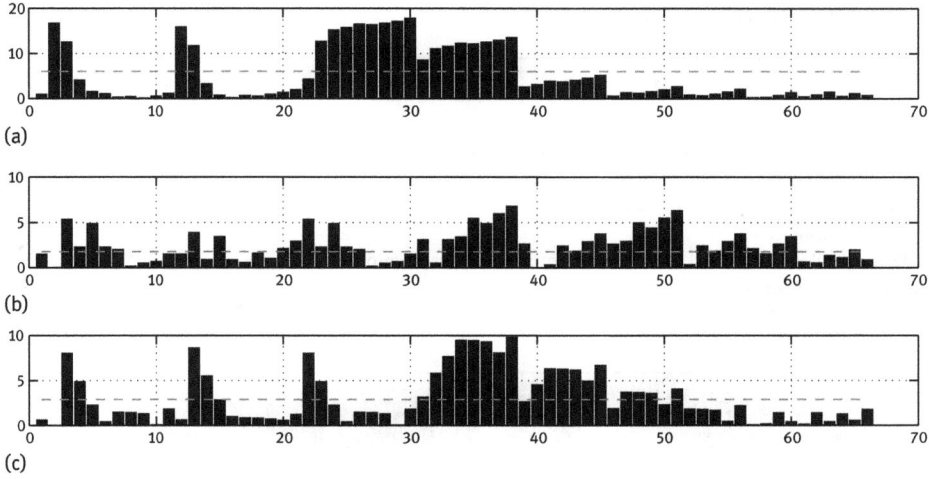

Fig. 12. Variance between details and threshold.

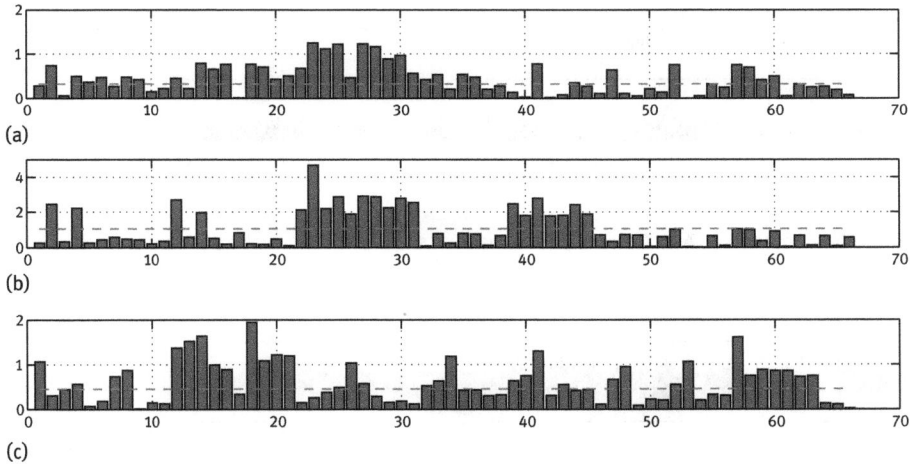

Fig. 13. Variance between principal components and threshold, (a) faut free case, (b) with fault 1, (c) with fault 2.

Let concider the binary $R_{nominal}$ (1 0 0), R_{fault1} (0 1 0), and R_{fault2} (0 0 1) output of the ANN to denote three classes of the residue.

The training parameters and the structure of the ANN used in this study are shown in Table 3.

Figure 14 and 15, present the outputs of the neural network for two cases (with no fault, with fault 1, with fault 2) using the Parseval's theorem.

Scenario 1 for fault 1 is proposed in this example. In this case, on the contrary of the training date, we change the period of the fault appearance. The fault takes place

Tab. 3. Ann architecture and training parameters.

Architecture	
Layer's number	3
Neurons on the layers	Input: 4, Hidden: 5, Output: 3
The initial weights	Random
Activation functions	Sigmoid
Training parameters	
Learning Rule	Levenberg-Marquart backpropagaion
Mean squared error	$10e^{-2}$

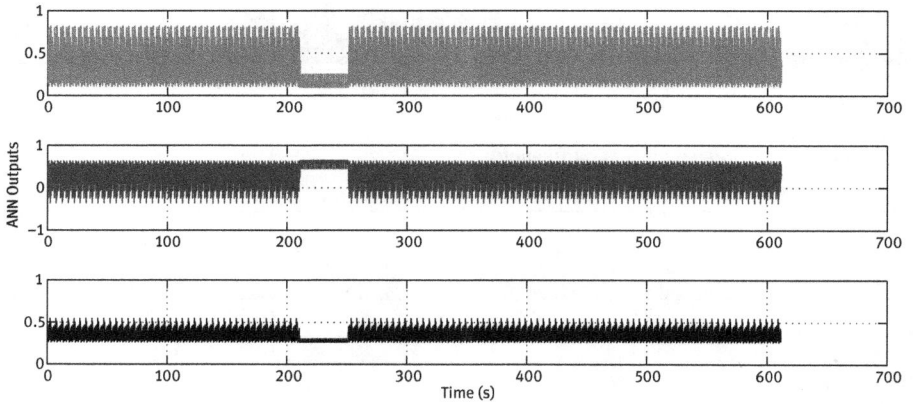

Fig. 14. Fault 1 detection: ANN Outputs for the residual signal with Parseval's theorem method.

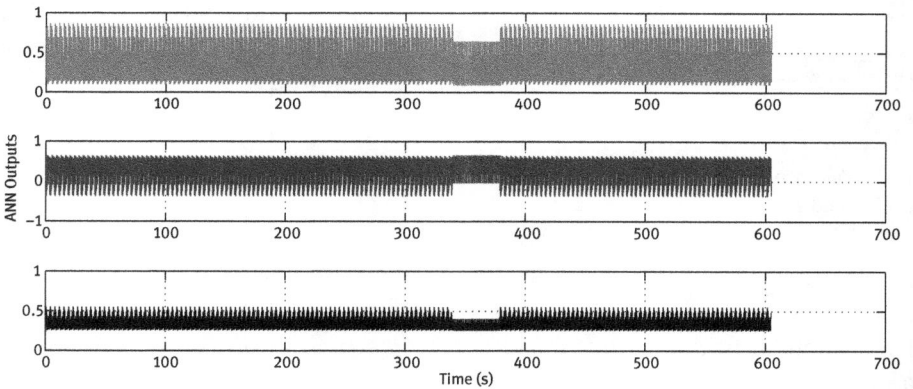

Fig. 15. Fault 2 detection: ANN Outputs for the residual signal with Parseval's theorem method.

between 210 s and 250 s. Figure 14 shows the three outputs of the network which given a binary value around (0,1,0) which correspond to R_{fault1} between the period 215 s and 250 s elsewhere the output is around (1,0,0).

Scenario 2 for fault 1 is proposed in this example. In this case, on the contrary of the training date, we change the period of the fault appearance. The fault takes place between 335 s and 380 s. Figure 15 shows the three outputs of the network which given a binary value around (0,0,1) which correspond to R_{fault1} between the period 335 s and 380 s elsewhere the output is around (1,0,0).

Figure 16, 17, present the outputs of the neural network for two cases (with no fault, with fault 1, with fault 2) using the PCA analysis.

Scenario 1 for fault 1 is proposed in this example. In this case, on the contrary of the training date, we change the period of the fault appearance. The fault takes place between 210 s and 250 s. Figure 16 shows the three outputs of the network which given

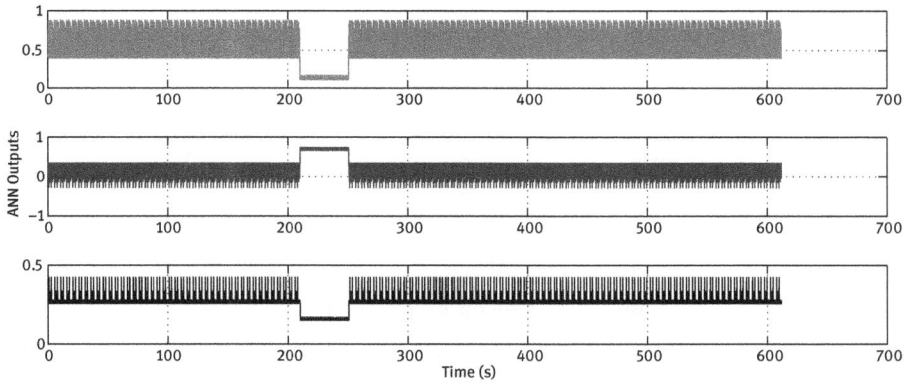

Fig. 16. Fault 1 detection: ANN outputs for the residual signal with PCA.

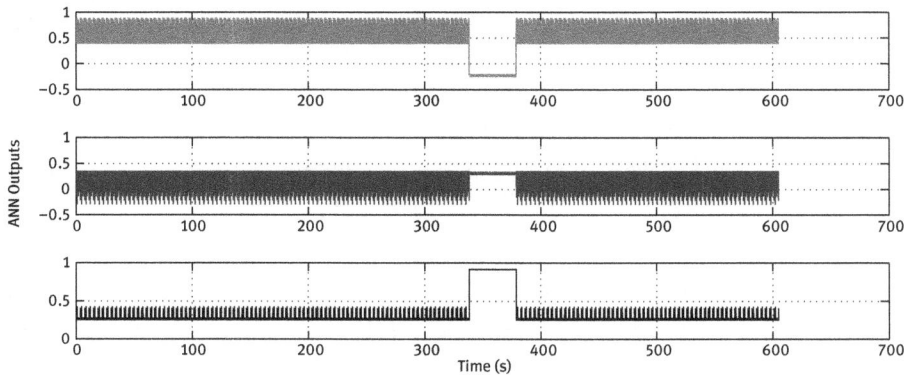

Fig. 17. Fault 2 detection: ANN outputs for the residual signal with PCA.

a binary value around (0,1,0) which correspond to R_{fault1} between the period 215 s and 250 s elsewhere the output is around (1,0,0).

Scenario 2 for fault 1 is proposed in this example. In this case, on the contrary of the training date, we change the period of the fault appearance. The fault takes place between 335 s and 380 s. Figure 17 shows the three outputs of the network which given a binary value around (0,0,1) which correspond to R_{fault1} between the period 335 s and 380 s elsewhere the output is around (1,0,0).

Using the suggested artificial neural network, system localize and identify the source of failures in the proposed cases. DWT/MRA and MLP architecture are useful for detecting and locating faulty in the monitored wind turbine system. In the present work, the accuracy rates achieved by the results showed that the proposed classifier based on PCA has the ability of recognizing and classifying residue signals efficiently compared to the Parseval's theorem method as two tools of feature selection. Details selected by the first method (PCA) are used better to detect and classify faults in comparison with the second method (Parseval's theorem energy). The use of the PCA enhances the chance of identifying faults especially when the system is very noisy.

4 Conclusion

In this paper, in order to detect and isolate faults in a wind turbine model, we employed the concept of DWT, MRA to decompose residual signals which are the inputs of the ANN classifier. Details given by this stage are extracted and selected using the energy Parseval's theorem and the PCA analysis. This two proposed hybrid methods of wavelet-based neural-network classifiers for residue classification can reduce the quantity of extracted features of generated residue signal without losing its property. Both not only can directly analyze residual signals but also save lots of computational cost, especially for large dimensional dimensional signals, due to the scale reduction of all the detailed selected at the price of increasing memory for the extracted features. Our simulation on a FAST wind turbine model indicate the Parseval's analysis as a feature extraction tool can consistently improve classification accuracy on residuals signals used but cannot necessarily work well compared to the corresponding PCA. The method proposed here for fault diagnosis and classification can be applied in the classification of other types of faults of the wind turbine system, or more generally in wind farm system. Once the ANNs are trained for different faults or scenarios, the classification process does not take much time. Therefore, an on-line version of the algorithm can be developed for real-time applications, possibly by implementing some of the processing in the hardware. As a future research direction, the design of ANN-based predictors can also be considered.

Acknowledgement: This work was supported by the Ministry of the Higher Education and Scientific Research in Tunisia.

Bibliography

[1]   I. Hwang, S. Kim, Y. Kim and C.E. Seah. A survey of fault detection, isolation, and reconfiguration methods. *IEEE Trans. on Control Systems Technology*, 18(3):636–653, 2010.

[2]   Y. Zhang and J. Jiang. Bibliographical review on reconfigurable fault-tolerant control systems. *Annual reviews in control*, 32(2):229–252, 2008.

[3]   J. Catalão, H. Pousinho and V. Mendes. Short-term wind power forecasting in portugal by neural networks and wavelet transform. *Renewable Energy*, 36(4):1245–1251, 2011.

[4]   K. Bhaskar and S. Singh. Awnn-assisted wind power forecasting using feed-forward neural network. *IEEE Trans. on Sustainable Energy*, 3(2):306–315, 2012.

[5]   B. Tang, W. Liu and T. Song. Wind turbine fault diagnosis based on morlet wavelet transformation and wigner-ville distribution. *Renewable Energy*, 35(12):2862–2866, 2010.

[6]   S. Lesecq, S. Gentil and I. Fagarasan. Fault isolation based on wavelet transform. *J. of Control Engineering and Applied Informatics*, 9(3–4):51–58, 2007.

[7]   F. Mo and W. Kinsner. Wavelet modelling of transients in power systems. *Conf. on Communications, Power and Computing*, WESCANEX'97, :132–137, 1997.

[8]   A. Gaouda, M. Salama, M. Sultan, A. Chikhani, et al., Power quality detection and classification using wavelet-multiresolution signal decomposition, IEEE Transactions on Power Delivery, 14(4):1469–1476, 1999.

[9]   J. Chung, E.J. Powers, W.M. Grady and S.C. Bhatt. Power disturbance classifier using a rule-based method and wavelet packet-based hidden markov model. *IEEE Trans. on Power Delivery*, 17(1):233–241, 2002.

[10] E. Schmitt, P. Idowu and A. Morales. Applications of wavelets in induction machine fault detection/aplicaciones de wavelets en la detección de fallas de máquinas de inducción. *Ingeniare: Revista Chilena de Ingenieria* 18(2):158, 2010.

[11] T. Bakir, B. Boussaid, P.F. Odgaard, M. Abdelkrim and C. Aubrun. WANN fault detection and isolation approach design for a fast wind turbine benchmark model. *Proc. of Engineering and Technology* (PET), :675–682, 2016.

[12] D.K. Sondhiya, A.K. Gwal, S. Verma and S.K. Kasde. Implementation of wavelet-based neural network for the detection of very low frequency (vlf) whistlers transients. 40th *COSPAR Scientific Assembly*, 40, 2014).

[13] H. Bendjama, D. Idiou, K. Gherfi and Y. Laib. Selection of wavelet decomposition levels for vibration monitoring of rotating machinery. 9th *Int. Conf. Advanced Eng. Comp. Applications in Sci.*, ADVCOMP, 2015.

[14] L. Placca, R. Kouta, D. Candusso, J.-F. Blachot and W. Charon. Analysis of pem fuel cell experimental data using principal component analysis and multi linear regression. *Int. J. of Hydrogen Energy*, 35(10):4582–4591, 2010.

[15] A. Nowicki, M. Grochowski and K. Duzinkiewicz. Data-driven models for fault detection using kernel pca: A water distribution system case study. *Int. J. of Applied Mathematics and Computer Science* 22(4):939–949, 2012.

[16] I. Daubechies. *Ten lectures on wavelets*. 61, CBMS-NSF Regional Conf. Series in Applied Mathematics, Society for Industrial and Applied Mathematics, 1992.

[17] H. Bendjama and M.S. Boucherit. Wavelets and principal component analysis method for vibration monitoring of rotating machinery. *J. of Theoretical and Applied Mechanics* 54(2):659–670, 2016.

[18] J.E. Dayhoff and J.M. DeLeo. Artificial neural networks. *Cancer*, 91:(S8):1615–1635, 2001.

[19] W. Musial, S. Butterfield and A. Boone. Feasibility of floating platform systems for wind turbines. 23rd *ASME Wind Energy Symp.*, Reno, NV, 2004.

[20] S. Butterfield, W. Musial and G. Scott. Definition of a 5-mw reference wind turbine for offshore system development. *Colorado: National Renewable Energy Laboratory*, 2009.

[21] P.F. Odgaard and K.E. Johnson. Wind turbine fault detection and fault tolerant control-an enhanced benchmark challenge. *IEEE American Control Conference*, (ACC), :4447–4452, 2013.

[22] P.F. Odgaard, J. Stoustrup, R. Nielsen and C. Damgaard. Observer based detection of sensor faults in wind turbines. *European Wind Energy Conference*, Marsielle, France, :4421–4430, 2009.

[23] T. Bakir, B. Boussaid, R. Hamdaoui, M. Abdelkrim and C. Aubrun. Qualitative diagnosis of wind turbine system based on wavelet transform, :1–6, 2015.

[24] T. Bakir, B. Boussaid, R. Hamdaoui, M. Abdelkrim and C. Aubrun. Fault detection in wind turbine system using wavelet transform: Multi-resolution analysis. *Int. Multi-Conf. on Systems, Signals and Devices*, :405–410, 2015.

[25] S. Simani. Application of a data-driven fuzzy control design to a wind turbine benchmark mode. *Advances in Fuzzy Systems*, 1, 2012.

[26] S. Simani, S. Farsoni and P. Castaldi. Fault diagnosis of a wind turbine benchmark via identified fuzzy models. *IEEE Trans. on Industrial Electronics*, 62(6):3775–3782, 2015.

Biographies

Tahani Bakir was born in Nefta, Tunisia, in 1986. She graduated from the National Engineering school of Gabes, Tunisia, in 2010. She is obtained her master degree in Automation and Intelligent Techniques in February 2012 from the National Engineering School of Gabes (Tunisia). Actually she is preparing her PHD degree. She is a member of the Modeling, Analysis and Control of Systems (MACS) research laboratory. Her research interests are in fault diagnosis and isolation in nonlinear system. Their application domains are mainly renewable energy (Wind turbine system).

Boussaid Boumedyen born in 1972 in Tunisia, he received a Ph. D. in control engineering in 2011 from Henri Poincaré University and from the National School of Engineers of Tunis, and an engineering degree in 1997 in electrical engineering from the National School of Engineers of Tunis. Since 1999 he has been an assistant professor in the Department of Electrical Engineering at the High Institute of Technology, University of Gabs. He is currently a member of the Research Centre for Automatic Control of Nancy (CRAN) and the Research Unit on Modeling Analysis and Control of Systems (MACS). His research interests focus on diagnosis and fault tolerant control areas with application to wind turbines.

Peter Fogh Odgaard received the M. Sc. degree in electrical engineering and the Ph. D. degree in control engineering from Aalborg University, Aalborg, Denmark, in 2001 and 2004, respectively. After Dr. Odgaard graduated he has worked at Aalborg University, kk-electronic a/s and Vestas Wind Systems. In December 2016, he joined Goldwind Energy as a Senior Control Engineer. He has authored more than 90 peer-reviewed conference and journal papers, and holds six patents and patent applications. His current research interests include control and fault diagnosis of wind turbines and wind farms. Dr. Odgaard has served as: a member of IFAC Technical Committee SAFEPROCES and PES Systems since 2011, Co-Chair of the Danish CSS/RAS joint chapter, IPC member of a number international conferences, special issue editor for international journals, and an Associate Editor at IEEE Transactions on Control Systems Technology since January 2017.

Mohamed Naceur Abdelkarim was born in Tunisia in 1958. He obtained a Diploma in Technical Sciences in 1980, his Master Degree in Control in 1981 from the ENSET school of Tunis (Tunisia), and his PhD in Control in 1985 and the Doctorate in Sciences Degree (Electrical engineering) in 2003 from the ENIT School of Tunis. He is a Professor at the Electrical Engineering Department (Control) of the National Engineering School of Gabes (Tunisia) and he is manager of the Modeling, Analysis and Control of Systems (MACS) laboratory.

Christophe Aubrun received a Ph. D. in control engineering from Henri Poincare University, France, in 1992. He is currently a member of the Research Centre for Automatic Control of Nancy (CRAN). Since 2005 he has been a professor in the Department of Electrical Engineering at the Institute of Technology, University of Lorraine. He has been involved in many projects with industry as well as European ones. His research interests lie in complex systems diagnosis and fault tolerant control areas with particular applications to water treatment processes and networked control systems.

B. Strauß and A. Lindemann

Implementation of a Temperature Measurement Method for Condition Monitoring of IGBT Converter Modules in Online-Mode

Abstract: The first part of this contribution, up to and including section "Accuracy analysis of the measurement concept", reproduces the results taken from the original paper published at the "International Multi-Conference on Systems, Signals and Devices 2016" in Leipzig, [10]. The second part adds results, which were presented in a submission at the "PCIM 2016" in Nuremberg, [12].

A measurement concept to determine junction temperatures based on a temperature sensitive electrical parameter (TSEP) of IGBTs will be described. For this purpose, the relationship between the used TSEP and the junction temperature is derived using semiconductor physical laws. Subsequently, the defined measurement concept will be described with respect to the function and its verification. In the context of a correlation and an accuracy analysis, a suitable calibration approach will be developed. Further, an integration of the measurement system into a three-phase converter application is described. In addition to results of a calibration procedure, the connection of the measurement system to the power circuits and a centralized data acquisition will be discussed.

Keywords: Junction Temperature, TSEP, IGBT, Threshold Voltage

1 Introduction

Power electronic modules are exposed to very high thermal loads during normal operation. This is in particular critical in case of operating conditions with low output frequencies of the respective converter. At higher temperature fluctuations over long time periods, the changing interlaminar thermomechanical stresses lead to wearout along the thermal path between chip and heat sink. This results in an increase of the junction temperature. Finally, various failure mechanisms may lead to the destruction of power electronic modules, [1], [2], [3]. Thus, it is needful to obtain information about the current health condition of the investigated converter module.

It is a challenge to monitor the junction temperature as precisely as possible and with a sufficiently high measurement sample rate. Besides simulating the thermal

B. Strauß and A. Lindemann: B. Strauß, Otto-von-Guericke University of Magdeburg, Germany, email: bastian.strauss@ovgu.de, A. Lindemann, Otto-von-Guericke University of Magdeburg, Germany, email: andreas.lindemann@ovgu.de

De Gruyter Oldenbourg, ASSD – Advances in Systems, Signals and Devices, Volume 7, 2018, pp. 179–200.
https://doi.org/10.1515/9783110470529-011

behavior [4], [5], [16], [17], [19], [20], measuring the module temperatures by negative temperature coefficient sensors (NTC-sensors) [6], [13] or the case temperatures [18], it is possible to determine the junction temperature by using temperature sensitive electrical parameters (TSEP), [21], [22], [23], [24], [25]. Initial feasibility studies have shown that the threshold voltage V_{th} of a semiconductor device is a suitable TSEP to determine T_j [8], [9]. This parameter has the advantage of a nearly linear behavior over the whole temperature range. Furthermore, the measurement of this TSEP is performed already during the currentless state of the respective semiconductor switch, so interferences by high rates of current changes are avoided. The resulting analog output voltage of a suitable measuring circuit corresponds to the threshold voltage V_{vth} of the respective semiconductor switch. Afterwards this output voltage is available for further signal processing as well as for a central evaluation by a higher-level control unit [12]. Finally, the individual junction temperatures of the IGBTs can be calculated using their approximated calibration curves.

2 Threshold voltage as a temperature sensitive electrical parameter [10]

The threshold voltage of an IGBT or MOSFET describes the gate-emitter voltage or the gate-source voltage at the point of time when the charge carrier inversion and thus the switching-on of the power semiconductor device begins. This happens, when the induced charge in the channel compensates the doping and thus Q_s becomes positive.

$$Q_s = C_{ox} \cdot (V_G - V_{th}) \tag{1}$$

In a first approximation, the temperature dependent threshold voltage can be described by a linear equation.

$$V_{th}(T_j) = V_{th,0} + \alpha \cdot (T_j - T_{j,0}) \tag{2}$$

Alternatively, with respect to [1] and [14] the theoretical relationship between threshold voltage and junction temperature may also be determined by using the laws of semiconductor physics.

$$V_{th}(T_j) = V_{FB} + 2\Phi_B + \frac{\sqrt{2 \cdot \varepsilon_{Si} \cdot q \cdot N_A \cdot (2\Phi_B)}}{C_{ox}} \tag{3}$$

where ε_{Si} is the permittivity of silicon, N_A the concentration of acceptors, q the elementary charge, Φ_B the intrinsic potential of the doped silicon chip as well as the flat band voltage V_{FB}. Here, the term flat band voltage describes a voltage of the gate electrode which still leads only to a flat energy band within the pn junction. During the switching-on process of a semiconductor, the energy levels of the conduction and the valence band are already very close to each other at this time. This state corresponds to a flat energy band gap. The above mentioned parameters are given by the following expressions:

$$\Phi_B(T_j) = \frac{\kappa \cdot T_j}{q} \ln\left[\frac{N_A}{n_i(T_j)}\right] \qquad V_{FB} = \Phi_{ms} - \frac{Q_f + Q_m + Q_{ot}}{C_{ox}} \qquad (4)$$

$$\Phi_{ms} = -\frac{\kappa \cdot T_j}{q}\left[\ln\left(\frac{N_A}{n_i}\right) + \ln\left(\frac{N_D}{n_i}\right)\right] \qquad (5)$$

Here N_D defines the concentration of donors and κ the Boltzmann constant. Q_f is the fixed oxide charge, Q_m the charge of mobile ions and Q_{ot} the intrinsic charge within the oxide. With respect to the intrinsic carrier density n_i, the electron density n as well as the hole density p at the temperature T_j are defined as follows.

$$n_i(T_j) = \sqrt{n(T_j) \cdot p(T_j)} \qquad (6)$$

$$n(T_j) = N_c(T_j) \cdot e^{\left(-\frac{W_c - \mu}{\kappa \cdot T_j}\right)} \qquad p(T_j) = N_v(T_j) \cdot e^{\left(+\frac{W_v - \mu}{\kappa \cdot T_j}\right)} \qquad (7)$$

W_v and W_c describe the energy levels of the conduction band and the valence band. For silicon, the state densities N_c and N_v of the conduction band and the valence band are defined in [14] by:

$$N_c(T_j) = 12 \cdot \left(\frac{2 \cdot \pi \cdot m_n \cdot \kappa \cdot T_j}{h^2}\right)^{3/2} \qquad N_v(T_j) = 2 \cdot \left(\frac{2 \cdot \pi \cdot m_p \cdot \kappa \cdot T_j}{h^2}\right)^{3/2} \qquad (8)$$

The parameters m_n and m_p are the effective masses of electrons and holes, h is the Planck constant. So the theoretical relationship between threshold voltage V_{th} and the absolute junction temperature T_j is calculable by

$$V_{th}(T_j) = \frac{\kappa}{q} \cdot \ln\left(\frac{N_A}{N_D}\right) \cdot T_j - \frac{Q_f + Q_m + Q_{ot}}{C_{ox}} + \frac{\sqrt{4 \cdot \varepsilon_{Si} \cdot N_A \cdot \kappa}}{C_{ox}} \cdot \sqrt{\ln\left[\frac{N_A}{n_i(T_j)}\right] \cdot T_j} \qquad (9)$$

In figure 1 the theoretical characteristic of the threshold voltage is documented based on equation (9) (blue) and for a simplified case (red).

It is visible that the threshold voltage V_{th} is dependent solely on one external physical variable, the junction temperature T_j. This behavior is obviously defined by the doping and thus mainly determined by the manufacturing process. For the documented temperature behaviour of the threshold voltage an exemplary but realistic doping configuration was used. Furthermore, a linear approximation showed, that a linearisation of equation (9) leads to a comparable temperature characteristic of $V_{th}(T_j)$. Thus, equation (2) is also applicable for junction temperature measurements.

In order to determine the junction temperature by the auxiliary variable V_{th}, a measurement concept had to be defined which will be described in the following section. Contrary to the theoretical relationship between threshold voltage V_{th} and

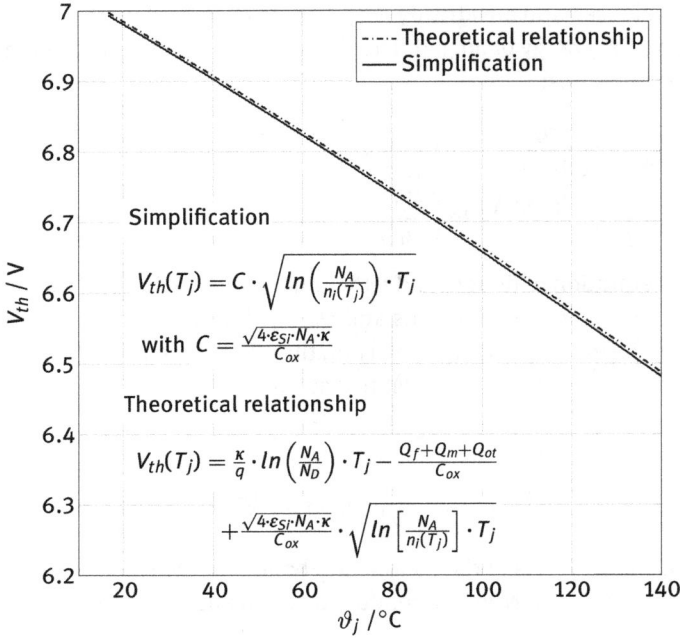

Fig. 1. Theoretical and simplified relationship between threshold voltage and junction temperature; with $T_j = \left(\frac{\vartheta_j}{°C} + 273\right) K$.

junction temperature ϑ_j the term **virtual** is used for measurements of these physical variables. Firstly, by indirect measurement methods for determining junction temperatures, it is not possible to measure the exact and theoretical value of ϑ_j. Secondly, by additional voltage drops across the gate circuit caused by gate currents through internal gate resistors and parasitic inductances, the measured threshold voltage differs from the theoretical value.

3 Measurement concept and function validation

To measure the virtual threshold voltage of an IGBT during the turn-on process as accurately as possible [10], it is necessary to generate a measuring trigger pulse V_{trig} over a defined period of time, see figure 2 and 3.

The scanning of the IGBT-side gate-emitter voltage begins when its value exceeds a minimum level of above 2 V. For this purpose, a time mask defines the time range when the measurement should be performed. This time mask is represented by the voltage signal V_{time}. The end of the sampling period will be defined by a trigger

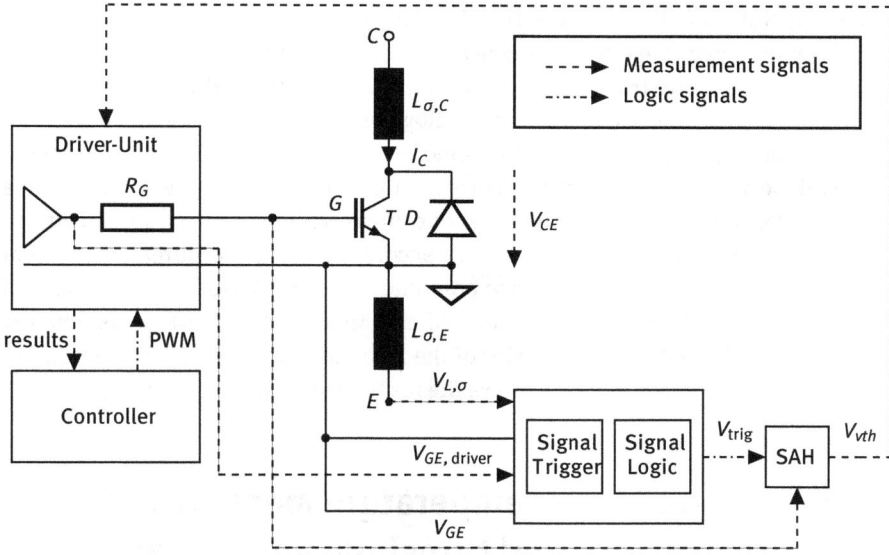

Fig. 2. Measurement concept for determining the junction temperature, based on [7].

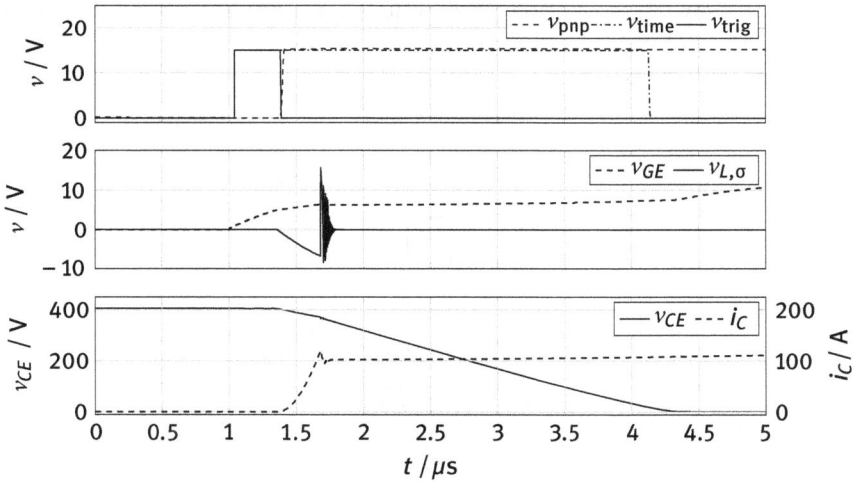

Fig. 3. Simulated behavior of a model according to the concept described in figure 2.

circuit which detects the beginning of the current flow through the respective power semiconductor switch. Thus, the turn-on trigger signal V_{pnp} is used to stop the sample cycle. Finally, by logical linking of V_{time} and V_{pnp} the corresponding measurement trigger pulse V_{trig} will be generated. If the measurement trigger pulse is HIGH, a

Sample-And-Hold circuit (SAH circuit) will sample the IGBT-side gate-emitter voltage V_{GE}. During this time, the output voltage of the SAH circuit follows the input voltage V_{GE}. If the measurement trigger pulse is LOW, the last sampled value is kept constant till the next sample cycle. Afterwards, the analog output voltage V_{vth} of the SAH circuit is available for further signal processing. Among other things, the V_{vth}-measurement values will be transformed into the corresponding junction temperatures by the central control unit. In figure 3, the logic and measurement signals are shown for a simulation model of a measuring circuit, according to the measurement concept illustrated above. Thereby the reference potential for each measurement and logical signal is given by the auxiliary emitter of the investigated IGBT (blue marked potential). To compare the real behavior of the measuring board with the simulation results depicted in figure 3, figure 4 shows the respective measurement signals.

4 Verification of IGBT temperature measurement using the TSEP virtual threshold voltage [10]

In this section some measurement results are presented to verify the measuring concept. In particular the correlation between junction temperature and threshold voltage is considered as well as systematic errors and statistic deviations of comparable measurement methods.

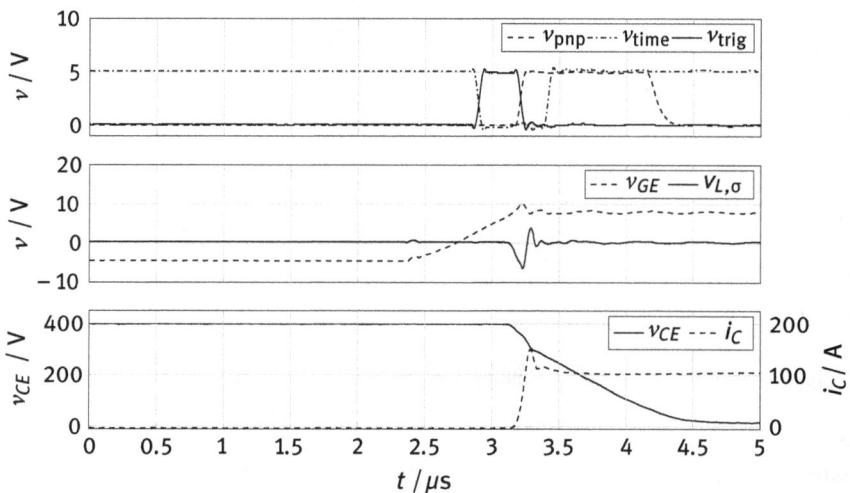

Fig. 4. Measured waveforms during turn-on of an IGBT.

Correlation analysis of $V_{vth}(\vartheta_{vj})$ [10]

The following statements are partly derived from [15]. The correlation analysis is used to determine the direction as well as the weight of relationship between at least two random variables X and Y. In the present case, these variables are defined on the one hand as the junction temperature $X \stackrel{\wedge}{=} \vartheta_{vj}$ and secondly as the threshold voltage $Y \stackrel{\wedge}{=} V_{vth}$. Subsequently, the standardized notations X and Y are used for describing both of the above mentioned random variables ϑ_{vj} and V_{vth}. The following calculations refer to the temperature characteristic of figure 7 in the following section. The arithmetic mean values of both random variables are calculable with:

$$\bar{X} = \frac{1}{n} \cdot \sum_{i=1}^{n} x_i = \underline{74.87°C} \qquad \bar{Y} = \frac{1}{n} \cdot \sum_{i=1}^{N} y_i = \underline{6.686V} \qquad (10)$$

The covariance S_{XY} defines a stochastic parameter which gives information about the relationship between two random variables. This parameter specifies whether with high values of X, the values of Y have rather low or also high magnitudes. The covariance is defined as

$$S_{XY} = \frac{1}{n-1} \cdot \sum_{i=1}^{n} \left(X_i - \bar{X}\right) \cdot \left(Y_i - \bar{Y}\right) = \underline{-5.21162K\ V} \qquad (11)$$

Both random variables have different ranges of values. So, the covariance is less meaningful. Thus, a covariance normalized to the respective range of values of both random variables is appropriate. This leads to the correlation coefficient ρ_{XY}. Further, the standard deviations of both considered random variables can be calculated with

$$\sigma_X = \sqrt{\frac{1}{n-1} \cdot \sum_{i=1}^{n} \left(X_i - \bar{X}\right)^2} = \underline{30,3791°C} \qquad (12)$$

$$\sigma_Y = \sqrt{\frac{1}{n-1} \cdot \sum_{i=1}^{n} \left(Y_i - \bar{Y}\right)^2} = \underline{0,1718V} \qquad (13)$$

According to [15], the correlation coefficient for two normally distributed random variables is determined by their standard deviations and the covariance with

$$\rho_{XY} = \frac{S_{XY}}{\sigma_X \cdot \sigma_Y} = \underline{\underline{-0,9986}} \qquad (14)$$

The range of values for the correlation coefficient ρ is defined from $-1 \leq \rho \leq 1$, [10]. In the present case it is a unidirectional correlation with a negative slope. This means that an increasing virtual junction temperature leads to a decrease of the measured virtual threshold voltage. For the above mentioned measurement results the correlation coefficient is approximately -1. In the present case a negative, very strong and thus linear correlation between both considered variables can be assumed. Thus, it is evident that a linearisation of the dependency between threshold voltage and junction temperature is acceptable and applicable. Another parameter for assessing the relationship between two random variables is the coefficient of determination. In the present case this parameter is defined by

$$B_{XY} = \rho_{XY}^2 = \underline{\underline{0,9972}} \tag{15}$$

Thus, 99:72 % of the dispersion of measured threshold voltages are describable by the scattering of temperature measurement within the linear relationship. In other words, for a temperature detection with low scatter of the measured values a correspondingly low scattering of voltage magnitudes results. This becomes also clear with respect to the statistical deviation in the following section.

Accuracy analysis of the measurement concept [10]

This section addresses the results of an accuracy analysis for the measuring concept to determine the junction temperature of a power semiconductor switch. Compared to the previous section, a new IGBT module was used to prevent incorrect measurements by previous damages.

For a suitable calibration procedure, the necessary determination of a reference junction temperature characteristic $V_{ref}(\vartheta_{vj,ref})$ was performed by using the NTC-sensors, which are already implemented in the power semiconductor module under investigation. Hence, for determining the real junction temperature, the NTC-sensors have to be calibrated as accurately as possible. To ensure a uniform temperature distribution within the module under test, a climatic chamber was used. A constant current source in temperature compensated mode was implemented to load the intrinsic NTC-sensors of the converter module. Across these NTC-sensors, the reference voltage $V_{ref}(\vartheta_{vj,ref})$ was recorded as an auxiliary variable for determining the reference temperature. The used test setup is given in figure 5. In order to formulate an analytical expression for calibrating the NTC-sensors, a polynomial approximation can be applied. To ensure a sufficiently high accuracy of the junction temperature measurement, the approximation of $\vartheta_{vj,ref}(V_{ref})$ was performed with a degree of 9. Thus, the i-th reference value of the virtual junction temperature can be calculated by

$$\vartheta_{vj,ref}(i) = A_0 + \sum_{k=1}^{9} A_k \cdot V_{ref}^k(i) \tag{16}$$

In figure 6, the calibration curve of one NTC-sensor is shown. Subsequently, the V_{vth}-measurement system had to be calibrated. For this purpose, a double pulse test setup was used. The electrical boundary conditions of the power circuit during the double pulse measurements were defined by $V_{DC} = 400V$ and $I_C = 130A$.

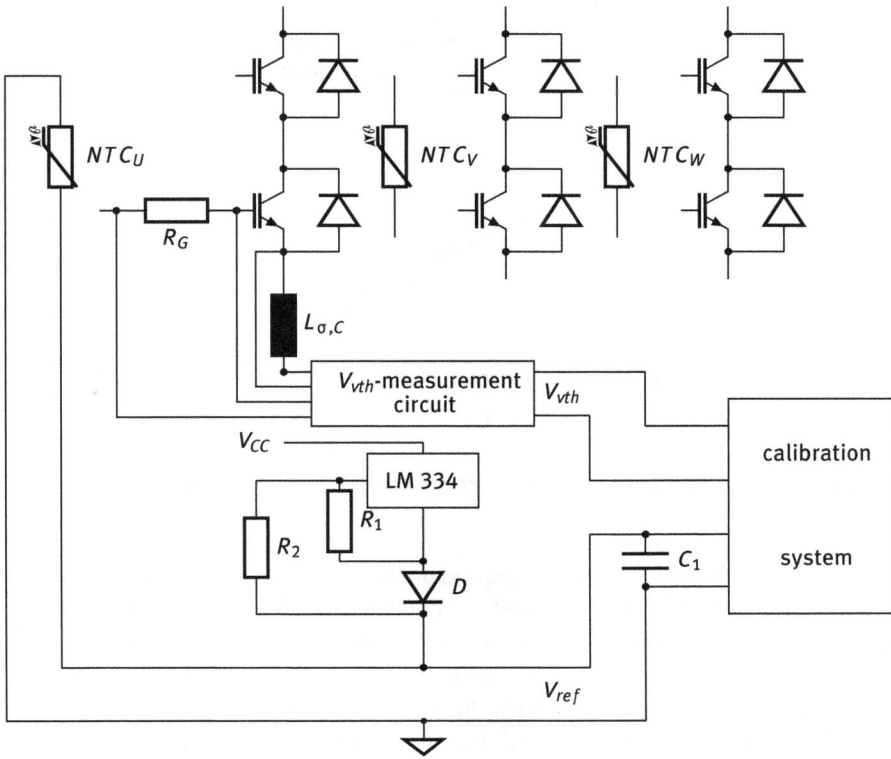

Fig. 5. Measurement system for determining $V_{vth,ref}(\vartheta_{vj,ref})$; exemplary for IGBT2.

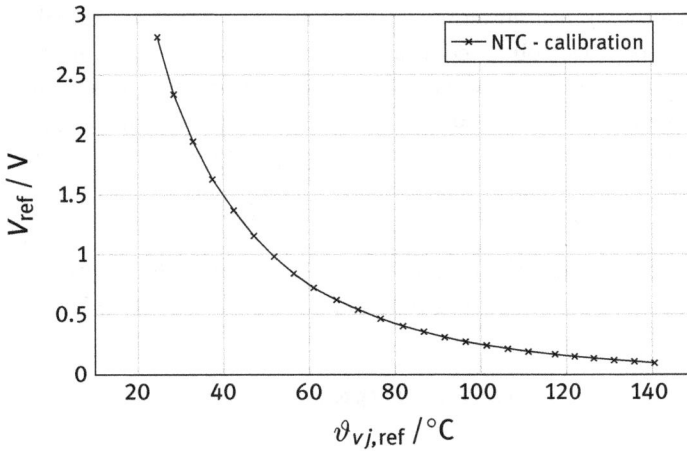

Fig. 6. Calibration curve of one NTC-sensor for determining the reference junction temperature in thermal steady state.

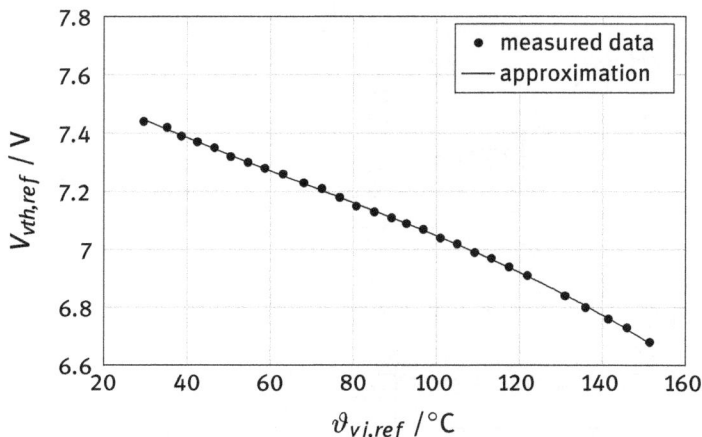

Fig. 7. Raw data and cubic approximation of the reference temperature characteristic of $V_{vth,ref}$.

The repetitions of double pulses were carried out after sufficiently long time intervals, to ensure that a self-heating by the load current can be neglected. Thus, each double pulse measurement was carried out during the thermal steady state. In this case the junction temperatures ϑ_{vj} correspond to the module temperatures measured by the intrinsic NTC-sensors. For this purpose the mentioned constant current source was also used in temperature compensated mode. The temperature setpoint was defined by using a controlled heating plate.

To generate the calibration curve $V_{vth}(\vartheta_{vj})$, the double pulse measurements were performed with temperature steps of 5K. The basis of an accuracy analysis is to determine the above mentioned reference measurement of $V_{vth,ref}(\vartheta_{vj,ref})$. The individual values of $V_{vth,ref}(\vartheta_{vj,ref})$ have been obtained by averaging the results of 50 individual measurements, each of them taken at the particular temperature under steady state conditions. The resulting analytical relationship between threshold voltage $V_{vth,ref}$ and junction temperature ϑ_{vj} can be expressed by a cubic approximation as follows

$$V_{vth,ref}(\vartheta_{vj}) = B_3 \cdot \vartheta_{vj}^3 + B_2 \cdot \vartheta_{vj}^2 + B_1 \cdot \vartheta_{vj} + B_0 \qquad (17)$$

with the coefficients B_0 to B_3 summarized in table 1. The temperature characteristic of $V_{vth,ref}(\vartheta_{vj,ref})$ as well as the respective approximation are depicted in figure 7.

Tab. 1. Coefficients of the polynomial approximation according to the reference measurement of $V_{vth}(\vartheta_{vj})$.

$B_3 \left[\frac{V}{K^3}\right]$	$B_2 \left[\frac{V}{K^2}\right]$	$B_1 \left[\frac{V}{K}\right]$	$B_0 [V]$
-1.6×10^{-7}	3.3×10^{-5}	-7.7×10^{-3}	7.65

Among other things, the results below are intended to show, that the range of values for the temperature characteristics of V_{vth} – determined by using the above mentioned reference measurement – was confirmed by comparable measurement procedures. Furthermore, because of the small inaccuracy it becomes clear that the above described reference measurement is best suited for a calibration process. Measurement procedure A was realized by using oscillograms of turn-on behavior of the corresponding IGBT. In the case of measurement procedure B, $V_{vth}(\vartheta_{vj})$ was determined by evaluating the analog output voltage of the V_{vth}-measurement circuit using the above mentioned oscillograms. An exemplary switching behavior is shown in figure 4. Measurement procedure C was performed using a precision multimeter for measuring the output voltage of the V_{vth}-measurement board during each temperature step. Detailed information about the different performed measurement procedures are given in [10]. In figure 8 the setups for these three measurement procedures are shown.

In comparison to the reference measurement results of $V_{vth,ref}(\vartheta_{vj,ref})$, the comparative temperature characteristics of procedure A, B and C are shown in figure 9. These comparative measurements were performed at the same time when the reference measurement using the calibration system was also carried out. Hence, all measured data are directly comparable with each other.

The comparison between reference characteristic and the results of measurement procedure C illustrates almost a coincidence of the measured temperature dependent V_{vth}-curves. Therefore, measurement procedure C can be adopted as the closest comparative measurement to the reference determination, already depicted in figure 7. So, these two procedures are predestined to calibrate the V_{vth}-measurement systems. In comparison to the precision multimeter, the characteristics measured with an oscilloscope are subject to a higher measuring deviation. Detailed information about the inaccuracy of the used measurement equipment are also given in [10]. As expected, a major reason for the greater deviations can be seen in the limited accuracy of the used oscilloscope. In figures 10 and 11 the absolute and relative errors are documented depending on the junction temperature. In the exemplary case of procedure A the absolute and relative errors are defined by:

$$ERR_A(\vartheta_{vj}) = V_{vth,A}(\vartheta_{vj}) - V_{vth,ref}(\vartheta_{vj}) \tag{18}$$

$$err_A(\vartheta_{vj}) = \frac{ERR_A(\vartheta_{vj})}{V_{vth,ref}(\vartheta_{vj})} \cdot 100\% \tag{19}$$

The absolute and relative errors in figure 10 and 11 illustrate again, that the inaccuracy of measurement procedure C is comparatively low. Among others, the corresponding variations in the absolute and relative errors of the measurements using an oscilloscope are primarily systematic. The averaged measurement error of procedure B has an almost constant behavior over the entire temperature range.

In this case, the correlating systematic errors can be considered to be nearly temperature independent. So, the measurement of $V_{vth}(\vartheta_{jv})$ based on an averaging

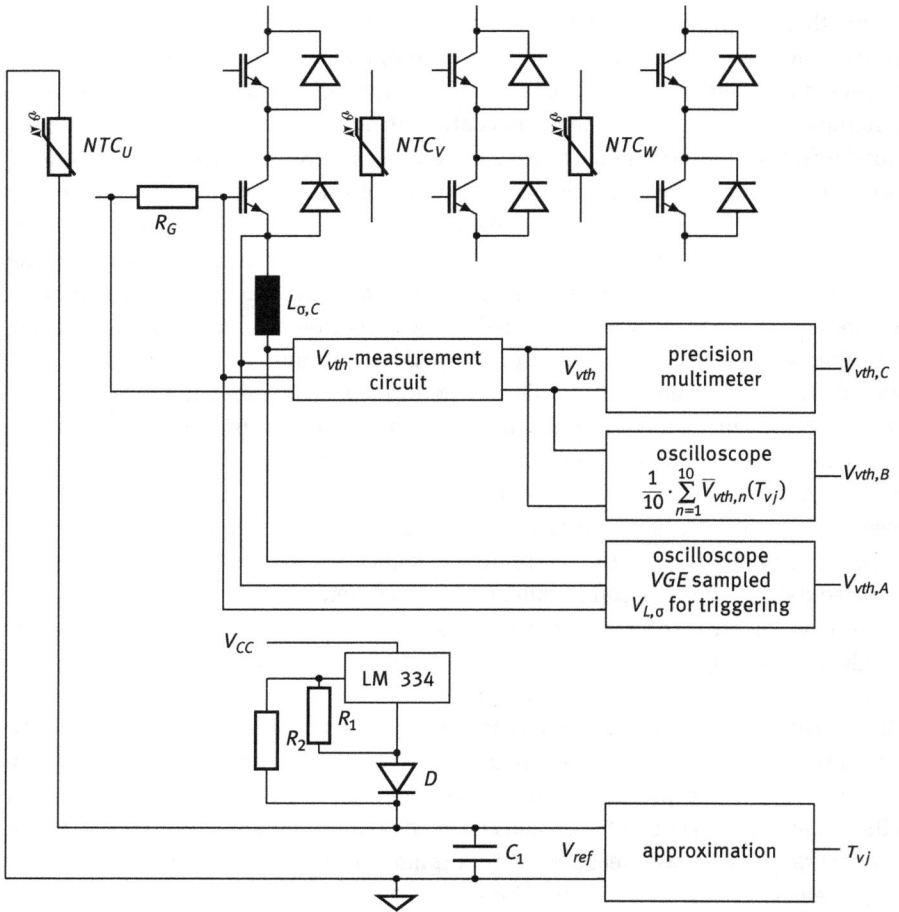

Fig. 8. Comparative measurement setups for determining $V_{vth}(\vartheta_{vj})$; exemplary for IGBT2.

of the measuring circuit output voltage using an oscilloscope, is influenced by an almost constant offset error. In contrast to the procedure B, measurement procedure A has an increasing averaged deviation from the reference measurement depending on the junction temperature. Here it is clear that the systematic error becomes larger with increasing junction temperature. The time delay of the measurement circuit should also be considered. On the one hand the time delay relates to the time block for generating the time mask based on the gate-emitter voltage and its $dv_{GE}(t)/dt$. On the other hand, the function block for triggering the beginning of turn-on is associated with a time delay between the $di_C(t)/dt$ and a response of the trigger block to the parasitic voltage $v_{L,\sigma}(t)$. Even through the logical linking of this trigger signal with the time mask, an additional time delay will be formed. In addition to

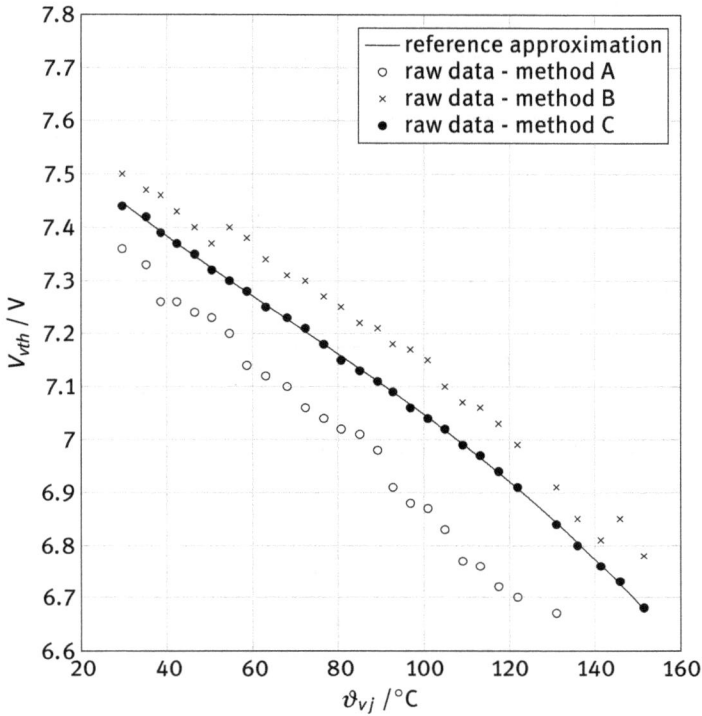

Fig. 9. Comparative temperature characteristics of V_{vth}.

the measuring inaccuracies of used equipment, the statistical error of the averaged measured variable will be considered more closely. The statistical error σ_m of an averaged random variable Y with n samples can be defined by:

$$\sigma_m(\bar{Y}) = \frac{\sigma_Y}{\sqrt{n}} \tag{20}$$

For the measurement procedures A, B and C, in figure 12 the statistical errors are shown depending on the reference junction temperature. Similar to the absolute and relative error depending on the reference temperature already shown in figure 10 and 11, the measurement procedures A and B show higher temperature independent statistical errors. Compared to the reference characteristic, the measurement procedure C was performed with less and only small statistical deviations.

For implementing a junction temperature measurement into a converter system, an auxiliary variable is required to determine the individual junction temperature indirectly. It is only meaningful to use a variable, which has a temperature independent systematic error and a small statistical deviation. This is the precondition to measure the junction temperature as accurately as possible by indirect measurement methods.

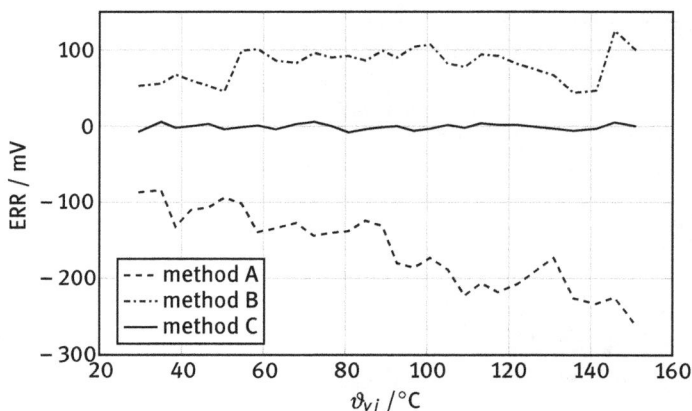

Fig. 10. Absolute error of measurement procedure A, B and C compared to the reference characteristic in figure 7.

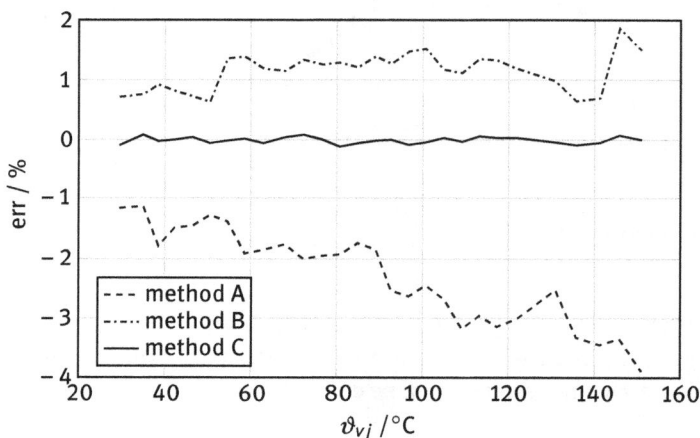

Fig. 11. Relative error of measurement procedure A, B and C compared to the reference characteristic in figure 7.

5 Implementation of the virtual threshold voltage Measurement concept into a power electronic application

In this section an integration of the developed measurement system into a power electronic application is described. As a device under test a standard three-phase converter module for drive systems of electric vehicles was used.

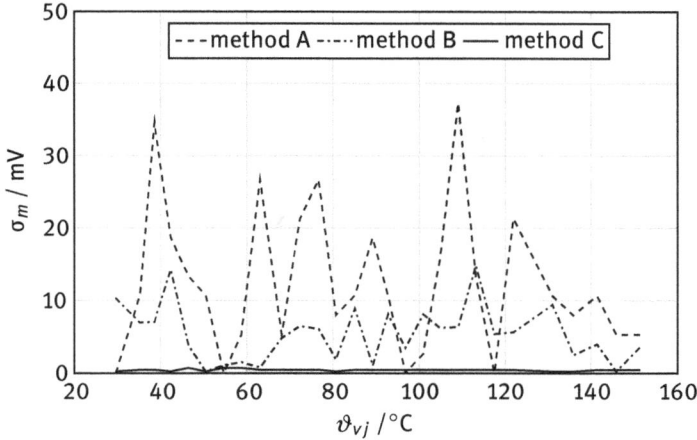

Fig. 12. Statistical errors of comparative measurement methods A, B and C related to the reference characteristic, depicted in figure 7.

5.1 Procedure for calibrating the threshold voltage measurement system to determine the junction temperature of IGBTs within a converter module [12]

A calibration during initial system setup helps to determine junction temperatures as accurately as possible although threshold voltages of the different IGBTs in the module package might slightly differ. A reference will again be required for such calibration. For this purpose, the above-described procedure to determine the reference temperature characteristics was also used for generating the temperature calibration curves of the investigated converter system. The respective characteristics of the intrisic NTC-sensors as well as the bottom side V_{vth}-measurement systems are depicted in figures 13 and 14. The resulting calibration curves of $V_{vth}(\vartheta_{vj})$ can be expressed by equation (17). The polynomial coefficients for approximating the threshold voltage characteristics of the investigated IGBTs are summarized in table 2. As mentioned above and with respect to [10] and [11], $V_{vth}(\vartheta_{vj})$ shows a nearly linear behavior. This becomes also clear by the low polynomial coefficients B_2 to B_3 in table 2. The coefficient B_1 describes the slope of the nearly linear characteristic. B_0 defines the offset at the temperature $\vartheta_{vj} = 0°C$.

In order to minimize computational resources of a centralized measurement recording, a linear or cubic approximation of the temperature characteristics would be sufficient. Due to the linear behavior, the calibration effort can also be reduced to a few temperature reference points. Furthermore, it is visible that each semiconductor chip is specified by its individual temperature characteristic $V_{vth}(\vartheta_{vj})$, as shown in

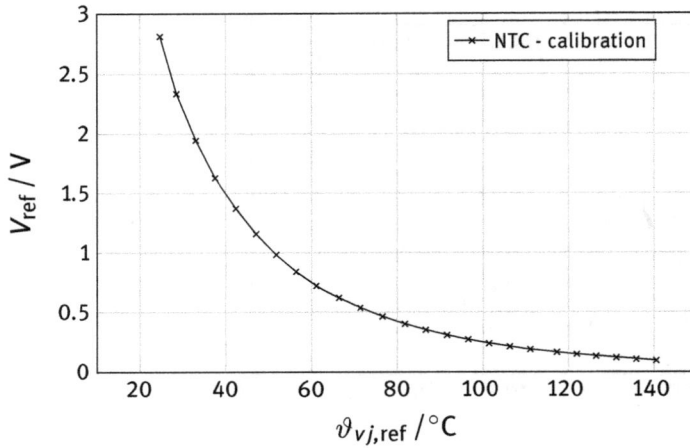

Fig. 13. Temperature characteristic of NTC-reference voltages; load current $0, 5$ mA.

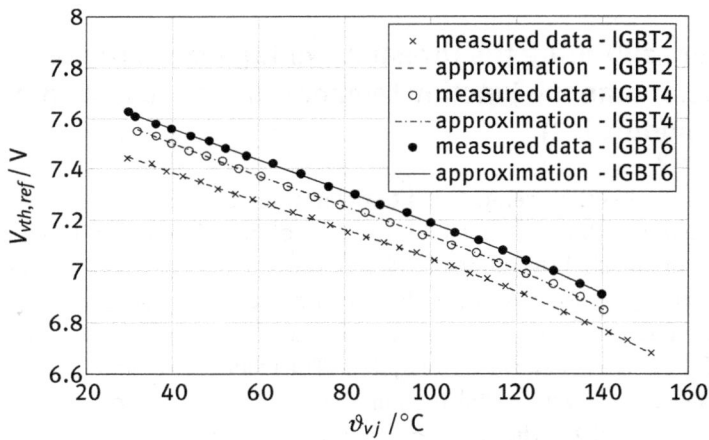

Fig. 14. Calibration curves of $V_{vth,ref}(\vartheta_{vj})$ for bottom side IGBTs of a converter module.

Tab. 2. Coefficients of polynomial approximations for calibration curves of $V_{vth}(\vartheta_{vj})$.

Semiconductor	$B_3\left[\frac{V}{K^3}\right]$	$B_2\left[\frac{V}{K^2}\right]$	$B_1\left[\frac{V}{K}\right]$	$B_0\,[V]$
IGBT2	-1.67×10^{-7}	3.41×10^{-5}	-7.78×10^{-3}	7.651
IGBT4	-2.19×10^{-7}	5.14×10^{-5}	-9.87×10^{-3}	7.826
IGBT6	-1.23×10^{-7}	2.70×10^{-5}	-8.01×10^{-3}	7.843

figure 14 and table 2. The slope of each curve is nearly the same, but the offset varies depending on the chip. For high accuracy of junction temperature measurement in a converter system, it is necessary to calibrate each individual semiconductor switch within the module.

5.2 Centralized recording of virtual threshold voltage measurement data in converter operation [12]

To realise an online monitoring of junction temperatures within a three-phase converter application, the calibration curves of six threshold voltage measurement systems can be implemented in a higher-level DSP. The measurement data will be transmitted via a suitable bus architecture across the potential differences between the power semiconductor switches and the controller. The centralized data acquisition as well as the final integration of six V_{vth}-measurement systems into a converter application is shown in this section.

Radiated and conducted disturbances will always be generated in pulsed mode operation of power electronic applications. In order to avoid consequent misinterpretations of the measured values – received by the higher-level DSP – a controller area network (CAN-bus system) is used for data transfer. The disturbances which are coupled in, will be minimized by a differential evaluation of the bus voltages. Effects on the information signals, caused by the CAN-bus impedance will also be eliminated. As a result, the information signals are only slightly disturbed and consist of clearly interpretable voltage levels, [12].

To implement the V_{vth}-measurement systems and the control unit into a centralized CAN-bus system, galvanic isolation between each bus component and the respective power semiconductor switch has to be realized. The integration of a V_{vth}-measurement circuit between a power semiconductor switch and a CAN-bus interface is depicted in figure 15. To digitize the analog output voltage of the V_{vth}-measurement circuit, a 12 Bit analog digital converter (ADC) is used. In addition to the ADC, the CAN-controller integrates the digitized values into standardized message frames, which will be transmitted via the CAN-bus subsequently. A CAN-transceiver guarantees the galvanic isolation between each IGBT with its individual measurement system and the centralized CAN-bus. The combination of measurement circuit, CAN-controller as well as CAN-transceiver forms a defined CAN node within the data communication system of the converter application.

Power supply of the respective driver can also be used for the measurement circuit and the CAN-controller inclusive the respective bus interface. The auxiliary emitter of the investigated IGBT defines the reference potential of each measurement circuit.

So, the equal constellation – consisting of driver unit, V_{vth}-measurement circuit as well as CAN interface stage – is implementable for each power semiconductor

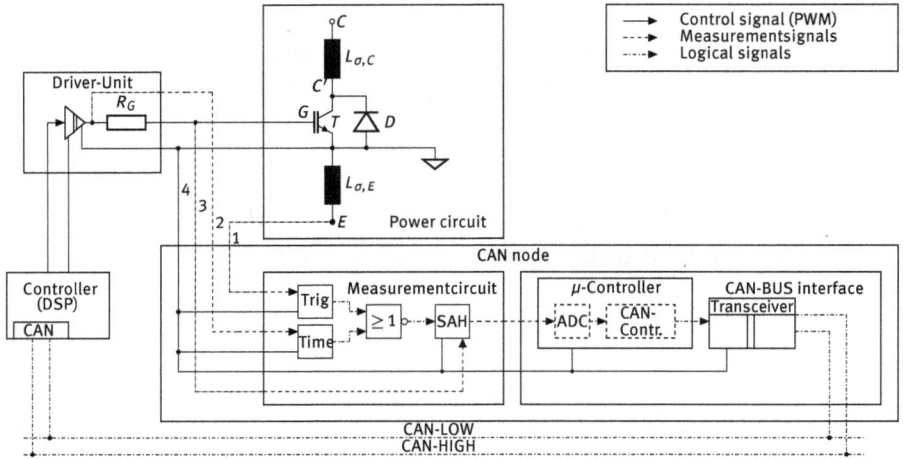

Fig. 15. Implementation of the V_{vth}-measurement circuit between a power semiconductor switch and the CAN-bus communication system.

switch in the same way. By the centralized DSP, the reference potential of the communication system – the digital ground – will be provided. The potential conditions of one CAN node are correspondingly depicted in figure 15 again.

In figure 16 the implementation of six V_{vth}-measurement systems in a three-phase converter application is illustrated. Here the defined signals and potential conditions are also visible. The communication system will be terminated by resistances of $R_{term} = 120\,\Omega$ at both ends of the CAN-bus. In interrupt mode, the DSP will acquire the centralized data recording during one measurement interval. To achieve a high time resolution of temperature measurement, the acquisition can be repeated in subsequent PWM periods. In this case, the maximum sample rate is limited by the

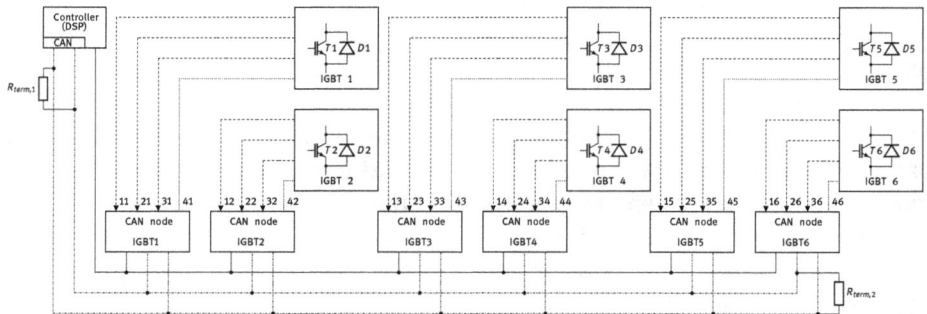

Fig. 16. Implementation of six V_{vth}-measurement systems into a three-phase converter application; measurement signals are defined by 1 to 4 for each individual IGBT, related to figure 15.

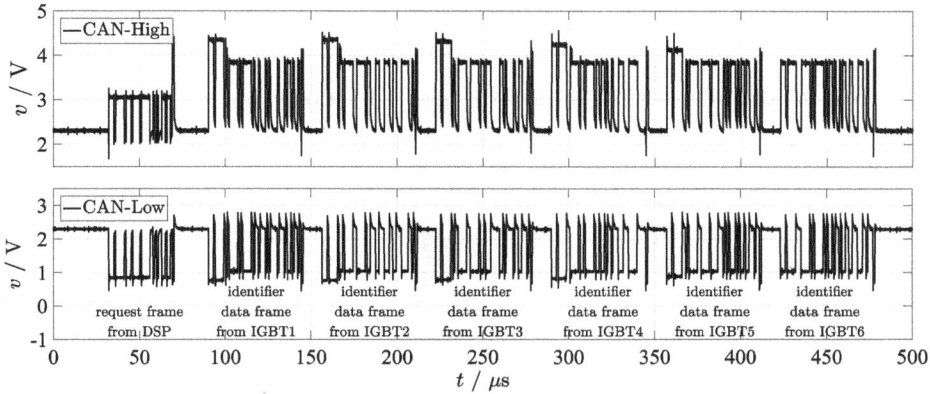

Fig. 17. Standard CAN-bus message frames for centralized recording of six digitized and averaged V_{vth}- measurements.

transmission speed of the CAN-bus and the computational load. The CAN-controller between the V_{vth}-measurement circuit and the CAN interface samples and digitizes the output values $V_{vth}(t)$ continuously. If required the DSP can also make the CAN-controller average digitized values of V_{vth} using a digital filter and store the respective mean value into the internal RAM. Afterwards, the DSP will request the mean value at each end of averaging period. In addition or alternatively, the CAN-controller could save the minimum value of digitized V_{vth}-measurement data. This way, the maximum junction temperature during one fundamental period of load current can be determined. The data transmission can be requested during the subsequent current less status of the investigated IGBT, if the respective free wheeling diode is in conduction mode.

Exemplary CAN-bus data frames for recording of six digitized and averaged threshold voltage measurement values are depicted in figure 17. Headed by a request frame, the DSP initiates a data transfer. As a serial response, the data frames of each individual V_{vth}-measurement system follow the request. Thereby, the individual objects of a CAN-bus are uniquely defined by identifier frames, which preceed each data frame. The high speed CAN-bus can be used with a maximum baudrate of $1\,\mathrm{MBit/sec}$. So the data transfer of six individual measurement values is realizable during a total time of $500\,\mu\mathrm{s}$. Depending on the computational load of the central DSP, every $0,5\,\mathrm{ms}$ the individual measurement values of junction temperatures can be updated. Standard power electronic modules are characterized by thermal time constants in the range of a second. So, for most three-phase IGBT converter applications, this time resolution will be sufficient, [10].

6 Conclusion

For a centralized and indirect determination of IGBT junction temperatures within a three-phase converter system of medium power class by using the virtual threshold voltage as an auxiliary variable, a measurement concept was implemented successfully. For this purpose the theoretical correlation between the threshold voltage V_{th} and the junction temperature ϑ_j has been derived. A calibration method to determine the temperature characteristic $V_{vth}(\vartheta_{vj})$ was developed. The linear temperature characteristic of the auxiliary variable V_{vth} used to determine the junction temperature ϑ_{vj} allows an efficient calibration procedure adjusting a few reference points. A cubic approximation of this temperature characteristic leads to a suitable calibration curve.

By the developed ϑ_{vj}-measurement method, the IGBT temperatures can be monitored. If one or more junction temperatures of the observed power semiconductor switches within a converter module reach their nominal values, a safety shut down of the system is possible. Furthermore, the long term analysis of junction temperatures can be used for investigations about influences of ageing mechanisms on ϑ_{vj}. Therefore, an approach for life time prediction of converter modules can be developed by using the junction temperature monitoring.

Bibliography

[1] J. Lutz, H. Schlangenotto, U. Scheuermann and R. D. Doncker. *Semiconductor Power Devices - Physics, Characteristics, Reliability. Springer Verlag*, 2011.

[2] K.B. Pedersen and K. Pedersen. Bond wire lift-off in IGBT modules due to thermomechanical induced stress. 3rd *IEEE Int. Symp. on Power Electronics for Distributed Generation Systems*, (PEDG), June, 2012.

[3] A. Middendorf. Lebensdauerprognostik unter Berücksichtigung realer Belastungen am Beispiel von Bondverbindungen bei thermomechanischen Wechselbeanspruchungen. *Diss., TU Berlin*, 2009.

[4] P.M. Igic, P.A. Mawby and M.S. Towers. Physics-based dynamic electro-thermal models of power bipolar devices (PiN diode and IGBT). 13th *Int. Symp. on Power Semiconductor Devices & ICs.*, IPSD '01, 2001.

[5] C. Yun, P. Malberti, M. Ciappa and W. Fichtner. Thermal component model for electrothermal analysis of IGBT module systems. *IEEE Trans. on Advanced Packaging*, 24(3):401–406, 2001.

[6] B. Ji, V. Pickert and B. Zahawi. In-situ Bond Wire and Solder Layer Health Monitoring Circuit for IGBT Power Modules. *Integrated Power Electronics Systems*, (CIPS) 2012.

[7] I. Bahun, N. Čobanov, Ž and Jakopović. Real-Time Measurement of IGBTs Operating Temperature. *Automatika*, 52(4), 2011.

[8] J.A. Butrón Ccoa, B. Strauß, G. Mitic and A. Lindemann. Investigation of Temperature Sensitive Electrical Parameters for Power Semiconductors (IGBT) in Real-time Applications. *PCIM 2014 Europe, Nuremberg*, 2014.

[9] B. Strauß and A. Lindermann. Indirect measurement of junction temperature for condition monitoring of power semiconductor devices during operation. *PCIM Europe*, Nuremberg, 2015.

[10] B. Strauß and A. Lindemann. Measuring the junction temperature of an IGBT using its threshold voltage as a temperature sensitive electrical parameter (TSEP). *Int. Multi-Conf. on Systems, Signals and Devices*, Leipzig, Germany, 2016.

[11] B. Strauß and A. Lindemann. Measurement of the junction temperature during operation of a drive converter. 13th *Braunschweiger Symp. Hybrid- und Elektrofahrzeuge, Braunschweig*, 2016.

[12] B. Strauß and A. Lindemann. Integration of a measurement circuit to determine junction temperatures of IGBTs in a three-phase converter. *PCIM Europe*, Nuremberg, 2016.

[13] A. Wintrich, U. Nicolai, W. Tursky and T. Reimann. Applikationshandbuch Leistungshalbleiter. *SEMIKRON Int. GmbH*, 2015.

[14] S.M. Sze. *Semiconductor Devices Physics and Technology*, 2nd Edition. John Wiley & Sons. Inc., 2002.

[15] L. León and U. Kiencke. *Messtechnik – Systemtheorie für Ingenieure und Informatiker*. Springer-Verlag, 2011.

[16] Z. Luo, H. Ahn and M.A.E. Nokali. A Thermal Model for Insulated Gate Bipolar Transistor Module. *IEEE Trans. on Power Electronics*, 19(4):902–907, 2004.

[17] B. Du, J.L. Hudgins, E. Santi, A.T. Bryant, P.R. Palmer and H.A. Mantooth. Transient Electrothermal Simulation of Power Semiconductor Devices. *IEEE IEEE Trans. on Power Electronics*, 25(1):237–248, 2010.

[18] Z. Wang, B. Tian, W. Qiao and L. Qu. Real-Time Aging Monitoring for IGBT Modules Using Case Temperature. *IEEE Trans. on Industrial Electronics*, 63(2):1168–1178, 2016.

[19] Z. Wang and W. Qiao. A Physics-Based Improved Cauer-Type Thermal Equivalent Circuit for IGBT Modules. *IEEE IEEE Trans. on Power Electronics*, 31(10):6781–6786, 2016.

[20] Z. Wang and W. Qiao. A Temperature-Dependent Thermal Model of IGBT Modules Suitable for Circuit-Level Simulation. *IEEE Trans. on Industry Applications*, 52(4):3306–3314, 2016.

[21] L. Dupont and Y. Avenas. Preliminary Evaluation of Thermo-Sensitive Electrical Parameters Based on the Forward Voltage for Online Chip Temperature Measurements of IGBT Devices. *IEEE Trans. on Industry Applications*, 51(6):4688–4698, 2015.

[22] Z. Xu, F. Xu and F. Wang. Junction Temperature Measurement of IGBTs Using Short-Circuit Current as a Temperature-Sensitive Electrical Parameter for Converter Prototype Evaluation. *IEEE Trans. on Industrial Electronics* 62(6):3419–3429, 2015.

[23] L. Dupont, Y. Avenas and P.-O. Jeannin. Comparison of Junction Temperature Evaluations in a Power IGBT Module Using an IR Camera and Three Thermosensitive Electrical Parameters. *IEEE Trans. on Industry Applications*, 49(4):1599–1608, 2013.

[24] Y. Avenas, L. Dupont and Z. Khatir. Temperature Measurement of Power Semiconductor Devices by Thermo-Sensitive Electrical Parameters, A Review. *IEEE IEEE Trans. on Power Electronics*, 27(6):3081–3092, 2012.

[25] H. Luo, Y. Chen, P. Sun, W. Li and X. He. Junction Temperature Extraction Approach With Turn-Off Delay Time for High-Voltage High-Power IGBT Modules. *IEEE IEEE Trans. on Power Electronics*, 31(7):5122–5132, 2016.

Biographies

Bastian Strauß Dipl.-Ing. Bastian Strauß works as a doctoral candidate at Otto-von-Guericke-Universität Magdeburg in Germany. He holds a degree in electrical engineering with specialisation in electrical energy since 2009. Until 2012, he worked as a measurement engineer for Matho Energiemanagement GmbH, which is specialized in industrial power distribution systems. Currently, he works as a researcher at the Chair for Power Electronics. Here, he focuses on procedures for online monitoring of converters during operation. A further research area is the electromobility, especially the required power electronics as well as the respective peripheral components. He is a member of the IEEE.

Andreas Lindemann Dr.-Ing. Andreas Lindemann holds a doctoral degree as electrical engineer. Having worked for ten years in power semiconductor industry, he since 2004 has been professor for power electronics at Otto-von-Guericke-Universität Magdeburg in Germany. His reserach is focused on the usage of novel power semiconductor devices in circuits and systems for various modern applications, considering special aspects like electromagnetic compatibility and reliability. He is volunteering in IEEE Power Electronics Society and VDE-ETG and chairman or committee member of several international power electronics conferences.